Sustainable Development Goals Series

The **Sustainable Development Goals Series** is Springer Nature's inaugural cross-imprint book series that addresses and supports the United Nations' seventeen Sustainable Development Goals. The series fosters comprehensive research focused on these global targets and endeavours to address some of society's greatest grand challenges. The SDGs are inherently multidisciplinary, and they bring people working across different fields together and working towards a common goal. In this spirit, the Sustainable Development Goals series is the first at Springer Nature to publish books under both the Springer and Palgrave Macmillan imprints, bringing the strengths of our imprints together.

The Sustainable Development Goals Series is organized into eighteen subseries: one subseries based around each of the seventeen respective Sustainable Development Goals, and an eighteenth subseries, "Connecting the Goals", which serves as a home for volumes addressing multiple goals or studying the SDGs as a whole. Each subseries is guided by an expert Subseries Advisor with years or decades of experience studying and addressing core components of their respective Goal.

The SDG Series has a remit as broad as the SDGs themselves, and contributions are welcome from scientists, academics, policymakers, and researchers working in fields related to any of the seventeen goals. If you are interested in contributing a monograph or curated volume to the series, please contact the Publishers: Zachary Romano [Springer; zachary.romano@springer.com] and Rachael Ballard [Palgrave Macmillan; rachael.ballard@palgrave.com].

Tsukasa Nagamine
Misa Masuda • Kenji Irie
Editors

Plant Genetic Resources (PGR) for Sustainable Crop Production

SATREPS Projects Tackling Global Warming and Environmental Extremes in Developing Countries

Springer

Editors
Tsukasa Nagamine
Ex-National Agriculture and Food Research Organization (NARO)
Tsukuba, Ibaraki, Japan

Misa Masuda
University of Tsukuba
Tsukuba, Ibaraki, Japan

Kenji Irie
Tokyo University of Agriculture
Setagaya, Tokyo, Japan

ISSN 2523-3084 ISSN 2523-3092 (electronic)
Sustainable Development Goals Series
ISBN 978-981-96-7116-8 ISBN 978-981-96-7117-5 (eBook)
https://doi.org/10.1007/978-981-96-7117-5

This Springer imprint is published by the registered company Springer Nature Singapore Pte Ltd.
The registered company address is: 152 Beach Road, #21-01/04 Gateway East, Singapore 189721, Singapore

Acknowledgments

We specially appreciate to the Dr. M. Kokubun, the Professor Emeritus of Tohoku University and the Former Research Supervisor of SATREPS Program, for giving us opportunity and encouragement to edit research and development conducted in SATREPS projects focusing on the use of plant genetic resources.

Contents

Plant Genetic Resources Contributing to Global Warming Mitigation and Improvement of Crop Production Under Extreme Environments

1

Tsukasa Nagamine, Misa Masuda, and Kenji Irie

Abstract

Crop production is always suffered by socio-economical conditions such as increase of fuel prices and environmental conditions such as climate changes. Particularly recent global warming is suffering sustainable crop production in developing countries; thus, we have to provide countermeasures against high temperature disorder, desertification, and salinization of agriculture land. Among several scientific and technological measures, the use of plant genetic resources (PGR) to improve crops genetically is most reliable and cost-efficient in developing countries. The Science and Technology Research Partnership for Sustainable Development (SATREPS) program promotes international collaborative research to address these global issues. The program is structured as a collaboration between Japan Science and Technology Agency and the Japan International Cooperation Agency. This book provides applicable examples of PGR and its genetic information solving issues caused by such as global warming and extreme environmental issues in the projects and also to present achievements by the use of novel PGR and its information on crop production in developing countries. This book covers various genetic resources, such as rice, wheat, tomato, chili pepper, quinoa, forage crops, cassava, banana, and rubber tree. Science and technology achievements obtained using PGR in developing countries by the projects could be useful for the increase of agricultural production in developed countries.

Keywords

CGIAR genebanks · Crop production · Extreme environment · Global warming mitigation · Plant genetic resources · Sustainable development goals (SDGs)

1.1 Current Situation of World Crop Production and Impacts of Global Warming on Crop Production

The world's estimated population in 1950 was 250 million peoples, while it reached 8 billion by mid-November 2022 (UN 2024). The increase in crop production has exceeded this, with produc-

T. Nagamine (✉)
Ex-National Agriculture and Food Research Organization (NARO), Ibaraki, Japan

M. Masuda
University of Tsukuba, Tsukuba, Ibaraki, Japan

K. Irie
Tokyo University of Agriculture, Setagaya, Tokyo, Japan
e-mail: k3irie@nodai.ac.jp

T. Nagamine et al. (eds.), *Plant Genetic Resources (PGR) for Sustainable Crop Production*, Sustainable Development Goals Series, https://doi.org/10.1007/978-981-96-7117-5_1

tion in 2022 being approximately 3.5 times greater than in 1961. On the other hand, the area harvested for crops has only increased slightly, about 1.1 times compared to the same year (FAOSTAT 2024). In other words, it is no exaggeration to say that various technological innovations in the agricultural field, including the Green Revolution, have increased land productivity and supported the population growth to date.

In crop production, if a suitable cultivation environment, superior varieties, and appropriate cultivation management techniques are applied, maximum yields and crops of excellent quality can be expected to be obtained. However, as the world's population continues to grow and the demand for meat and dairy products increases due to the diversification of dietary habits, feed crop production will also expand. Furthermore, abnormal weather conditions such as high temperatures, droughts, and heavy rains associated with global warming pose a potential threat to food production, including crops, and there is a risk that the balance between food supply and demand will be disrupted. In fact, in recent years, in agricultural fields around world, high temperatures during growing season, droughts, poor crop production, and yields due to lack of irrigation water, and declining quality of grains and fruits due to the effects of global warming, new tropical- and subtropical-derived pests in mid- and high-latitude regions, soil erosion due to floods, desertification, and loss of large areas of the wildfires in large area have occurred frequently.

In crop production, it is expected that optimal cultivation conditions, good crop varieties, and appropriate cultivation management techniques will produce the highest yields and harvests of superior quality. Recently, however, there have been many problems in agricultural fields around the world that can be attributed to the effects of global warming: high temperatures during the growing season, lack of irrigation water, poor crop production and yields due to drought, poor grain and fruit quality, new tropical and subtropical pests in mid and high latitudes, soil erosion due to flooding, desertification, and the loss of agricultural land due to wildfire outbreaks has become a frequent occurrence. One of the measures to combat global warming, increased demand for bioethanol and biodiesel crops, may in turn put additional pressure on arable land and forests, as long as the total land mass remains constant.

The impact of global warming on crop production is more severe in developing countries than in developed countries (Cline 2007). It is becoming increasingly important to combat global warming in cooperation with developing countries by improving crop varieties and cultivation management techniques.

These activities to develop agricultural technologies to adapt to climate change are in line with Goal 2 of the Sustainable Development Goals (SDGs) (https://sdgs.un.org/goals) proposed by the United Nations: "zero hunger, achieve food security and improved nutrition, and promote sustainable agriculture."

1.2 SATREPS Program to Solve Global Issues

The effects of global warming are not confined to a single country but extend over a wide area. To address such global-scale issues that affect a wide area, it is necessary to establish and conduct international joint research that transcends national boundaries.

The Science and Technology Research Partnership for Sustainable Development (SATREPS) program is an international joint research project that aims to solve global-scale issues as part of "Science and Technology Diplomacy," in which science, technology, and diplomacy are linked and mutually developed. Global-scale issues such as climate change, food problems, natural disasters, and infectious diseases are becoming more complex and serious every year. In order to solve these issues, it is necessary to create science and technology innovations that transcend national boundaries, return the results of research to society, and develop human resources and research capabilities. (https://www.jst.go.jp/global/english/about/about.html).

Table 1.1 Major research theme in the research area of bioresources of SATREPS

	Major research theme
1	Research contributing to the sustainable production and utilization of plant resources
2	Research on evaluation and effective utilization of underutilized biological resources and biological functions
3	Research on the improvement of the environment in which bioresources are located
4	Research on mitigation of the effects of climate change on bioresource production
5	Research on production and utilization of new food resources (alternative protein foods (proteins derived from plants and animals), genome-edited foods, and cellular foods) and on environmental improvement

In the SATREPS program, researchers from Japan and developing countries will work together to address issues in the four research area of environment and energy, bioresources, disaster prevention, and infectious diseases and create new knowledge and technologies that can be used in the real world while taking local needs into account, thereby contributing to an international society aiming for sustainable development. In the research area of bioresources, we are working to contribute to an international society that aims for sustainable development.

In the research area of biological resources, research on sustainable production and utilization of biological resources will be conducted. Specifically, we will conduct the research area listed in the Table 1.1.

In the 16 years since the SATREPS program began, 48 projects have been conducted in the research area of bioresources, 30 of which are plant projects.

Many research and technological developments using plant genetic resources have been conducted under this program, and significant results have been achieved in combating global warming and improving crop production under extreme conditions. This book introduces the results of crop research and development using plant genetic resources conducted under the SATREPS program and looks forward to the future use of plant genetic resources.

The crops, countries, and other information presented in this book are listed in Table 1.2.

1.3 Plant Genetic Resources for Improving Crop Production in Extreme Environments

There are some areas in the world where crops must be cultivated in fields with extreme environmental conditions such as hot, dry, cold, overhumid, saline, oligotrophic soils, or at high latitudes.

On the African continent, weathered soils that have been exposed to rain, wind, and high temperatures for many years predominate, various nutrient deficiency other than nitrogen inhibit crop growth. Paddy soils in Madagascar have low phosphate content, resulting in poor early rice growth. In addition, farmers apply low amounts of fertilizers in conventional rice cultivation, resulting in low yields. In order to increase rice production in these soil environments, it is expected that the introduction of new crop varieties using genetic resources and effective cultivation techniques are effective.

In the Altiplano of Bolivia, there is no other crop that can be grown except quinoa (*Chenopodium quinoa*). However, recently, production has been unstable due to low yields caused by pre-harvest frost damage and mildew. Therefore, it is desirable to develop quinoa varieties that use genetic resources that have early maturity, which allows for earlier harvesting in the Altiplano, and large grain size, which contributes to higher yields.

1.4 Use of Wild Species and Traditional Knowledge

Some wild species grow (and survive) in environments such as saline soils and arid conditions and exhibit saline soil tolerance, drought tolerance, and pest resistance. These biological and non-biological resistances (tolerances) of wild species

Table 1.2 Crops, countries, traits targeted for improvement, and main plant genetic resources used in the research area of bioresources of SATREPS

Crop	Country	Targeted traits for improvement	Genetic resources used
Rice	Madagascar	Improved yield in low fertility soils	Phosphate uptake gene *Pup1*
Rice	Kenya	Yield, abiotic and abiotic stresses	IRRI collection, African wild rice
Wheat	Afghanistan	Poor environment tolerance	Indigenous Afghan varieties
Wheat	Sudan	High temperature tolerance	Multiple synthetic wheat derived populations using wheat (*Aegilops tauschii*), etc.
Wheat	India	Adding biological nitrification suppression	Intergeneric hybrids of wheat with wild relatives (*Leymus recemosus*)
Quinoa	Bolivia	Early maturing, large grain	Indigenous variety, closely related wild species
Tomato and chili pepper	Indonesia	Heat tolerant tomato and anthracnose resistant chili pepper	TILLING tomato lines, indigenous pepper varieties and TILLNG pepper lines
Forage grass	Mongolia	Rapid growing	*Chloris* wild species
Cassava	Cambodia and Vietnam	Cassava mosaic virus disease	CIAT collections
Banana	Peru	Banana fusarium wilt resistance	INIA collection
Rubber tree	Indonesia	Leaf blight resistance	Collection of the Indonesian Rubber Research Institute

are important for breeding new crop varieties that can be cultivated in poor environments. However, in highly developed crop varieties, the degree of genetic segregation increases in the hybrid progeny of wild and improved varieties, increasing the time required for selection and fixation and, in some cases, missing out on the superior characteristics of the improved varieties. For these reasons, it is considered desirable to avoid the direct use of wild genetic resources for breeding parents.

Sudan is the hottest and driest wheat growing region in the world. Existing wheat genetic resources are not sufficient to breed wheat varieties with stable yield and high quality in the country, and novel genetic resources that utilize wild species need to be developed. A multiple synthetic wheat derivative population has been developed using the closely related wild species of wheat (*Aegilops tauschii*). The use of wild genetic resources for breeding wheat varieties grown under hot and dry conditions is promising.

The short arm of the n-chromosome of wild species of wheat (*Leymus racemosus*) with high biological nitrification (BNI) is introduced into wheat variety in the project of India.

Chromosome fragment substitution lines of African wild rice *Oryza longistaminata* have been used as a genetic resource for blast resistance in rice breeding in Kenya. The efficacy of the wild species for blast resistance is discussed.

The traditional knowledge of the characteristics and medicinal properties of plant genetic resources that farmers have developed over many years of experience in collecting wild species and cultivating indigenous varieties can provide useful information for the improvement of crop varieties.

Quinoa is the only crop that can be grown in the Bolivian Altiplano, but it is sometimes unable to produce stable growth and yields due to global warming. Ajara (*Chenopodium quinoa* ssp. *melanospermum*), a closely related wild species of quinoa, is early maturity and can grow well and produce seeds in extreme environments. There is hope that Ajara can be used to breed early maturing quinoa varieties.

Among the forage grass species growing in poor-nutrient soils in Mongolia, a fast-growing wild species of *Chloris virgata* has been discovered based on nomadic observations and traditions, and a project has been initiated to use this genetic resource for forage grass breeding. Basic physiological, genetic, and nutritional analyses of rapid growth in oligotrophic soils have revealed useful genes, and functional analysis of these genes is underway.

1.5 Breeding of Vegetatively Propagated Crops Using Genetic Resources

This book reports on the research and development projects on three vegetatively propagated crop projects: cassava, banana, and rubber tree. The cassava project is concerned with Cassava Mosaic Virus disease (CMD) caused by Sri Lanka Cassava Mosaic virus, the banana project is concerned with banana fusarium wilt caused by the banana wilt fungus, and the rubber tree project is concerned with leaf blight caused by *Pestalotiopsis* spp. and other leaf blight pathogens. As with the three crops, the research goal of the project is to improve disease resistance, which is not a coincidence, but is something of a fate for vegetatively propagated crops that are cultivated as clones of the same genotype. The potential for improving disease resistance in vegetatively propagated crops using genetic resources will be discussed.

1.6 Importance of the CGIAR Genebank

According to the second edition of the World Food and Agriculture Plant Genetic Resources, in 2009 there were 1,750 collections worldwide, including the Consultative Group on International Agriculture Research (CGIAR) Institutions, of which 130 collections hold more than 10,000 accessions. The genetic resources preserved in these collections are expected to be used by breeders and researchers.

The Kenyan rice project has introduced genetic resources from the International Rice Research Institute (IRRI) Genebank to breed drought-tolerant, cold-tolerant, salt-tolerant, and blast-resistant varieties. Promising lines suitable for the Kenyan growing environment have been selected.

In the rice project in Madagascar, about 360 lines of indica-type rice varieties, including many from Aus rice and Madagascar, were introduced from IRRI's Genebank, where SNP information is publicly available, and GWAS analysis was conducted to elucidate for chromosomal regions that can produce high yields even in low-fertility soils in the region. The results of genome-wide association (GWAS) analysis using the rice core collection of single nucleotide polymorphism (SNP) information are expected to be promising.

Among nitrogen fertilizers applied in the field, excess nitrogen that is not absorbed by crops becomes nitrate nitrogen through nitrification by soil microorganisms, which pollutes groundwater. It is also returned to nitrous oxide nitrogen, which has a high greenhouse effect and is a cause of global warming. The wheat project in India will utilize the power of plants to suppress soil nitrification and improve crop productivity while improving the nitrogen cycle. Biological Nitrification Inhibition (BNI)-enabled international wheat lines were introduced from International Maize and Wheat Improvement Center (CIMMYT) for the project. It will be interesting to see how the cultivation methods using BNI-enabled wheat lines with reduced nitrogen fertilizer application in the field in India will produce growth and yield results.

Cassava SATREPS projects in Cambodia and Vietnam have introduced 53 lines of promising cassava breeding material from the Centro Internacional de Agricultura Tropical (CIAT) to breed CMD-resistant varieties.

The importance of the CGIAR Genebank in the utilization and conservation of plant genetic resources will be discussed.

References

Cline WR (2007) Global warming and agriculture: if steps are not taken to curb carbon emissions, agricultural productivity could fall dramatically, especially

in developing countries. Finance & Development, 23–27. March 2007
United Nation World Population Prospects (2024) Revision. https://population.un.org/wpp
FAOSTAT (2024) World food and agriculture statistical yearbook 2024. Food and Agriculture Organization. https://openknowledge.fao.org/

Enhancing Rice Yields with Limited Fertilizer Use and Poor Soil Nutrients in Madagascar: Dissemination of *Pup1* Variety and P-dipping Technique

2

Yasuhiro Tsujimoto

Abstract

In sub-Saharan Africa, rice production from smallholder farms remains extremely low due to limited fertilizer inputs and nutrient-deficient soils. Therefore, improving nutrient-use efficiency is a critical challenge for increasing rice yield in the region. In the project entitled "Breakthrough in Nutrient Use Efficiency for Rice by Genetic Improvement and Fertility Sensing Techniques in Africa," we aimed to overcome this challenge by developing new varieties with the *Pup1* locus, which enhances root development; promoting the P-dipping technique, which refers to the localized phosphorus application to seedling roots at transplanting; and utilizing rapid soil diagnosis methodologies to enable selective fertilizer applications. Furthermore, socioeconomic assessments have revealed that increasing rice yields improves farmers' cash revenue and nutritional status, particularly in terms of energy and micronutrients, by accelerating the purchase of nutritious food items (vegetables, fruits, and meat/fish) in Madagascar. The research output was disseminated to thousands of local farmers to help address food security issues in Madagascar. This study provides a practical example of a transdisciplinary approach for delivering research outputs to beneficiaries by involving researchers, policymakers, extension officers, and private entities, such as fertilizer dealers and seed producers.

Keywords

Madagascar · Nutrient use efficiency · P-dipping · Pup1 · Transdisciplinary approach

2.1 Introduction

Rice is the most rapidly expanding food commodity in sub-Saharan Africa (SSA). Rice consumption has tripled in the past 20 years, and currently, in SSA, rice is the second largest source of caloric intake after maize (FAOSTAT 2024). De Vos et al. (2023) estimated that rice consumption in SSA will double by 2050 relative to that in 2020, owing to the continuous increase in population and per capita consumption. Therefore, without a significant yield gain, the ever-growing demand for rice is anticipated to further increase reliance on expensive imports or cropland expansion (Saito et al. 2023). The latter scenario of excess cropland expansion can lead to environmental degradation, such as biodiversity and carbon loss (Alexander et al. 2023), as well as an

Y. Tsujimoto (✉)
Japan International Research Center for Agricultural Sciences, Tsukuba, Ibaraki, Japan
e-mail: tsujimotoy0365@jircas.go.jp

T. Nagamine et al. (eds.), *Plant Genetic Resources (PGR) for Sustainable Crop Production*, Sustainable Development Goals Series, https://doi.org/10.1007/978-981-96-7117-5_2

increase in methane emissions from flooded lowlands (Chen et al. 2024).

The current rice yield in the SSA is still 2.1 t ha^{-1}, which is far below the average of 5.0 t ha^{-1} in other regions (FAOSTAT 2024). One of the major constraints on rice yield is that most farmers cultivate rice with little mineral fertilizer input in highly weathered soils (Rakotoson et al. 2022; Saito et al. 2019; Tsujimoto et al. 2019). In SSA, the fertilizer inputs for agricultural production are merely 8.7 kg ha^{-1} of nitrogen (N) and 1.8 kg ha^{-1} of phosphorus (P), while the average values are 76.0 and 12.3 kg ha^{-1}, respectively, in the other regions of the world (FAOSTAT 2024). Our survey of 2,236 rice fields in the central highlands of Madagascar showed that the farmers applied mineral fertilizers to only 12% of their fields (Ozaki and Sakurai 2021). Moreover, a recent surge in fertilizer prices has made it more difficult for poor farmers to access mineral fertilizers. In 2022, the International Fertilizer Association estimated a 5% decline in global fertilizer consumption from 2021; however, this reduction rate in fertilizer consumption from 2021 to 2022 was most significant at 25% in SSA (World Bank 2023).

Subsidy programs geared toward lowering the net cost of fertilizers are effective means of increasing farmers' economic incentives to apply more fertilizers (Sanchez 2015). However, excessive subsidies can be a burden on the national economies of SSA countries; hence, the cost may outweigh the benefits over the long term, as funding for subsidies is typically directed away from other agricultural investments (e.g., research for innovations) that may eventually have greater impacts on agricultural development (Jayne et al. 2013). Additionally, excessive subsidies are considered to reduce nutrient use efficiencies by applying excessive amounts of fertilizers or by applying fertilizers to less fertilizer-responsive fields using less efficient techniques. The inappropriate use of fertilizers may cause excess nutrient outflow to the ecosystem and accelerate the depletion of finite nutrient resources, such as rock phosphates (Zhang et al. 2015; Zou et al. 2022). Studies have pointed out that subsidy programs may not favor poor farmers because they often cultivate less fertilizer-responsive soils with less efficient management practices (Marenya and Barret 2009; Ricker-Gilbert and Jayne 2012). Therefore, technical innovations are fundamental for sustainably increasing crop yields under limited fertilizer inputs and nutrient-deficient soils in SSA.

In the project titled "Breakthrough in Nutrient Use Efficiency for Rice by Genetic Improvement and Fertility Sensing Techniques in Africa" under the SATREPS program, we worked on the issue in a bilateral research collaboration between Japan and Madagascar. In Madagascar—one of the largest rice-producing countries in SSA—70% of the rice cultivation area is categorized as "irrigated" and mostly managed by smallholder farmers. Irrespective of the relatively favorable water conditions, the average yield was low at less than 3 t ha^{-1}. The restricted yield is primarily due to inadequate fertilizer application and poor nutrient supply capacity of highly weathered soils. For example, ferralsols spread across the central and eastern parts of the island, corresponding to the major rice-producing areas.

Deficiencies in nutrients such as phosphorus, nitrogen, sulfur, and silicate restrict rice productivity. The stagnation of rice cultivation, which affects more than half of the population, risks national food security and deprives poor farmers of the chance to increase their livelihood. This situation has prevented the country from escaping poverty and hunger: the country's extreme poverty rate of 81% (as of 2012, accessed at https://pip.worldbank.org/country-profiles/MDG) and hunger rate of 49% (FAO et al. 2024) are both far above the averages for SSA countries. Thus, Madagascar is a representative case, in which solutions for improving nutrient-use efficiency are key to increasing rice yields and contributing to the eradication of poverty and hunger. This project aimed to develop rice production techniques to achieve high yields under low-fertility conditions by optimizing fertilizer use suited to the soil nutrient characteristics of the field and creating new breeding lines with high nutrient use efficiency. We evaluated the impact of adopting such techniques on the income and nutritional status of local farmers.

2.2 Soil Diagnosis for Selective P Application

Soil nutrient diagnosis is a conventional approach that allows site-specific nutrient management practices and improves nutrient use efficiency. Previous studies have implied that a lack of information about the nutrients to be applied and the expected returns deprives farmers of the appropriate use of fertilizers (Abay et al. 2022; Van Campenhout 2021). These findings are particularly relevant for smallholder farmers in SSA because their fields are known to be highly heterogeneous in terms of soil fertility and responses to nutrient inputs, even within small distances, because of the influence of topography and past management practices (Kihara et al. 2016; Zingore et al. 2011; Tsujimoto et al. 2017). Heterogeneity in soil characteristics among neighboring fields was observed in the target regions of the central highlands of Madagascar (Nishigaki et al. 2019). Thus, site-specific information will undoubtedly improve productivity and nutrient use efficiency.

However, comprehensive soil analysis is costly and time-consuming, and such detailed information does not necessarily ensure effective fertilizer management. This is because the relationship between soil properties and crop responses to fertilizer input in the field is complex and not straightforward (Tsujimoto et al. 2019). This is particularly true for lowland rice production systems, because the biological and chemical reactions of soils are complex under flooded conditions. As a countermeasure, we focused on a single nutrient element, phosphorus (P), as the major yield-limiting factor for lowland rice production in the region, based on the results from repeated nutrient-omission trials in pots and in farmers' fields (Fig. 2.1).

Using the results of the aforementioned omission trials, we identified that the response of rice plants to P application could be estimated with the P-retention capacity of soils (Nishigaki et al. 2021). As illustrated in Fig. 2.2, the response to P application—the yield difference between NP- and N-applied plots—declined linearly in fields with higher P-retention capacity. The results

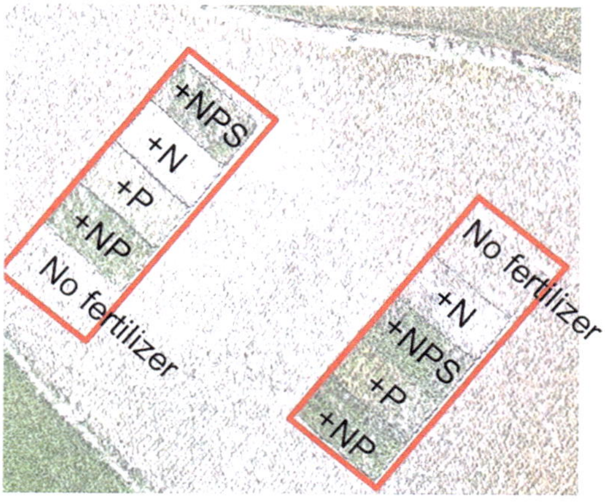

Fig. 2.1 On-farm nutrient-omission trials in the central highlands of Madagascar. The plant growth is severely reduced when phosphorus (P) is omitted, i.e., in the No fertilizer and +N plots. N, P, and S indicate nitrogen applications as urea, phosphorus as triple superphosphate, and sulfur as sodium sulfate

implied that the yield response to P application was greater at sites with higher elevations and cooler climates, which is attributable to the additional effect of P application to avoid low-temperature stresses by accelerating the maturity of rice plants. Although the importance of soil P-retention capacity for crop responses to P fertilizer has been addressed in a previous study (Nziguheba et al. 2016), our results are the first to demonstrate a quantitative relationship between soil P-retention capacity and yield responses to P application in lowland rice production in farmers' fields. According to the International Soil Reference and Information Center, it is estimated that more than 25% of SSA soils are classified as having "moderate to very high" P-retention capacities (Batjes 2011).

Furthermore, we revealed that the air-dried soil moisture content can predict the soil P-retention capacity across a range of soil types (Nishigaki et al. 2023). Soil moisture content can be measured with high reproducibility by placing soil samples (1–10 g) with a saturated salt solution inside a closed container as a moisture conditioning agent. This prediction methodology is easy to apply without laboratory facilities or chemical analysis and thus provides practical utility for agricultural extension officers to

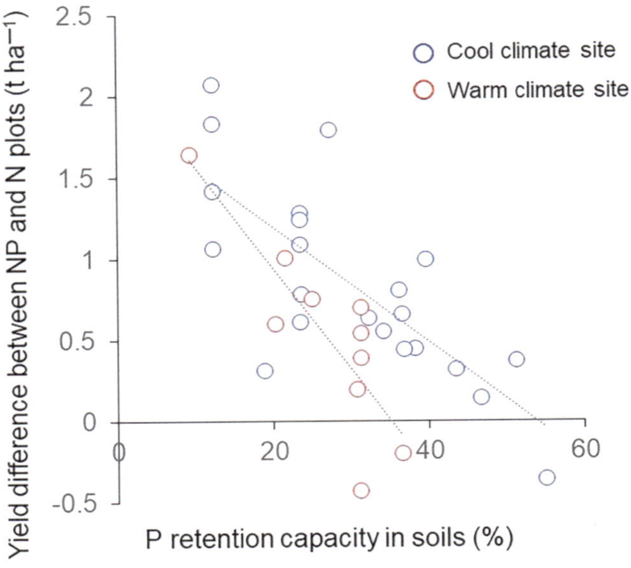

Fig. 2.2 Relationship between the response of rice yield to phosphorus (P) application and P-retention capacity in soils

identify fields responsive to P fertilization. Furthermore, using a randomized controlled trial of 70 farmers in 5 villages, our socioeconomic trials demonstrated the effectiveness of providing simple and plot-level soil information (Ozaki et al. 2024). In the trial, farmers who received information about the P deficiency status of their fields optimized fertilizer allocation and increased their rice yield and income at the household scale.

2.3 Localized Nutrient Application via P-dipping

Simple and reliable methodologies for soil diagnosis can help farmers judge where to apply P fertilizers more efficiently. Are the options in case fields judged to have a high P-retention capacity and are less responsive to P fertilizer applications? A conventional solution is to apply large amounts of P fertilizers above the P-retention capacity of soils, as is the case in volcanic soils in developed countries (Dahlgren et al. 2004). However, this option is not suitable to smallholder farmers in SSA and is less favorable in terms of environmental impacts and the finite nature of P fertilizer resources.

In this project, we developed a localized P application technique—P-dipping—which refers to the placement of P in the root zone by adhering P-enriched slurry (a mixture of P fertilizer and puddled paddy soils) to seedling roots at transplanting (Fig. 2.3). Through repeated pot and on-farm trials, this technique has been proven to be highly effective in increasing lowland rice yields, irrespective of the P-retention capacity of soils.

First, we examined the optimal P concentration and sources for developing the slurry and the optimal duration for dipping the seedling roots in the slurry. The effect of such a localized application increases by concentrating more nutrients near the roots; however, this increases the risk of salt stress, known as fertilizer burning. Experimental results suggested that the P concentration at 1.8–2.8% (w/w) using triple superphosphate (TSP) is optimal to accelerate plant growth while avoiding the risk of fertilizer burning (Oo et al. 2020a). We observed that the risk of fertilizer burning increased when old seedlings or high temperatures were used during transplanting (Oo and Tsujimoto 2022; Rakotoarisoa et al. 2023). The key function of this technique is not to develop P-enriched seedlings but rather to transfer ample amounts of P to the main field in the form of slurry adhered to seedling roots at transplanting (Oo et al. 2020a). Therefore, the

Fig. 2.3 Illustration of the P-dipping technique (on the left) and its effect on rice growth on high-P-retention soils (on the right)

duration of dipping in the P-enriched slurry can be instant, whereas a prolonged dipping duration of > 4 h may increase the risk of salt stress to the seedlings (Oo et al. 2020a). With this manipulation, we identified that P-dipping creates soluble P hotspots near the root zone and develops vigorous surface root systems, which, in combination, facilitate P uptake by rice plants, even in high P-retention soils (Fig. 2.3; Oo et al. 2020b).

Subsequent on-farm trials confirmed that P-dipping can significantly increase both grain yield and applied P-use efficiency in the typical P-deficient lowlands of Madagascar (Rakotoarisoa et al. 2020; Tsujimoto et al. 2021; Oo et al. 2023). Rakotoson et al. (2022) summarized that P-dipping could achieve agronomic P-use efficiency (yield gain per unit of P applied) twice as much as conventional P application via broadcasting. P-dipping enables relatively large amounts of P input into the main field, whereby a positive P balance between external inputs and removal by harvesting can be retained. Thus, the risk of a negative P balance or P mining from the soil is of little concern with this technique, even with high P-use efficiency (Rakotoson et al. 2022). However, further empirical evidence is required to assess the long-term effects of phosphate dipping on grain yield and soil fertility.

Repeated on-farm trials demonstrated that the effect of P-dipping was particularly prominent in fields at high elevations and in late-transplanted fields at low elevations. This is because P-dipping alleviated P deficiency and late-season low-temperature stress by shortening the days to heading (Fig. 2.4a, c). With P-dipping, the growth duration of rice was shortened by a maximum of 1 month in the P-deficient lowlands of Madagascar, which provided a significant advantage for reducing the risks of biotic and abiotic stresses (Rakotoarisoa et al. 2020). Additionally, the P-dipping effect increased when the fields had erratic water levels after transplantation, which was associated with vigorous initial growth and avoidance of submergence stress (Oo et al. 2023; Fig. 2.4b, c). Both phosphorus deficiency and weather-induced stresses, such as low temperatures at the reproductive stage and flash flooding at the initial growth stage, are prevalent in lowland rice-producing areas in SSA (Ismail et al. 2013; van Oort 2018; Saito et al. 2019). Furthermore, the impacts of climate change on agricultural production and food security are expected to be most significant in SSA (Baptista et al. 2022), where farmers already face many weather-induced shocks owing to poor infrastructure. Therefore, P-dipping can be an attractive strategy for dually addressing improvements in P-use efficiency and adaptation to climate change for lowland rice production in SSA.

The effectiveness of this technique was verified under various edaphic climatic conditions and farming management practices in Madagascar. Technical training of P-dipping was provided to more than 3,000 local farmers by regional extension officers and the Japan International Cooperation Agency (JICA)

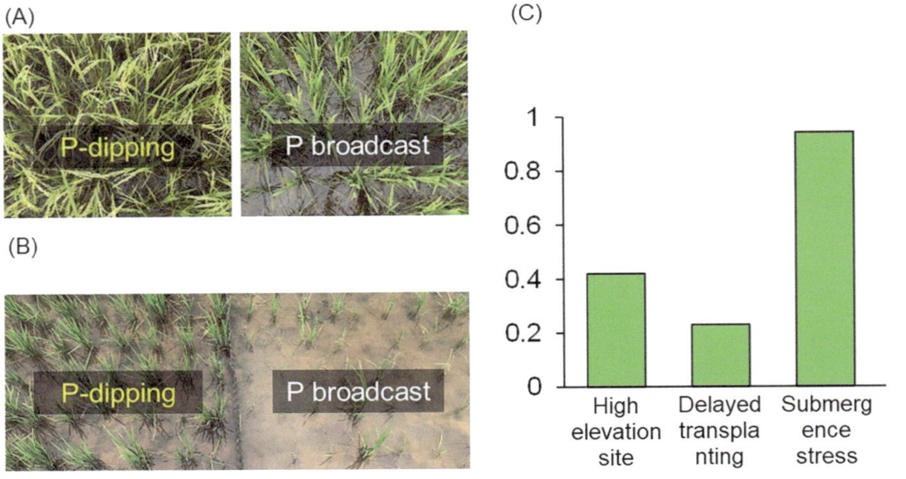

Fig. 2.4 Effect of P-dipping under a range of various environmental conditions. (**a**) Effect to shorten the growth duration. (**b**) Effect to accelerate initial growth and avoid submergence stress. (**c**) Partial regression coefficient of determinant factors for the P-dipping effect on rice yield

Fig. 2.5 A farmer adopting P-dipping (on the left) and a local shop selling small packages of triple superphosphate (TSP) for P-dipping (on the right)

technical cooperation project. Then, many farmers adopted the technique, increasing rice yields by 30%, from 3.7 t ha^{-1} to an average of 4.8 t ha^{-1} (Fig. 2.5). In response to the dissemination of this technique among farmers, private companies in Madagascar began selling fertilizers (triple superphosphate, TSP) for P-dipping. The anticipated increases in labor costs at transplantation and access to TSP are potential constraints for the further dissemination of the technique. In our survey of 962 farmers who practiced P-dipping across the central highlands of Madagascar, 32% cited the increased labor required for slurry preparation and for teaching the new method to transplant workers as a negative aspect of the technique. However, 95% of them were willing to continue the technique because of its simplicity and yield benefit. Further socioeconomic analyses or extension activities are needed to compare the benefits and additional labor costs for wider applicability of the technique.

2.4 Variety Development Under P-deficient Conditions

Genetic improvement is another strategic approach for improving nutrient use efficiency and increasing rice yields under low fertilizer input and nutrient-poor soil conditions. Root morphology is a key factor in exploring and acquiring nutrients from soils more efficiently. The importance of the root system is emphasized for P because it is relatively immobile in soils compared to other elements such as N and K (Campos et al. 2018; Wissuwa 2005). Wissuwa and Ae (2001) demonstrated that genotypic variation in P uptake in rice can be largely explained by root size when cultivated under P-deficient conditions. The research group identified a major quantitative trait locus (QTL) for phosphorus deficiency tolerance, *Pup1*, in the traditional Aus type variety Kasalath (Wissuwa et al. 1998; Wissuwa and Ae 2001). Subsequently, the locus was sequenced, revealing the presence of a *Pup1*-specific protein kinase gene, *PSTOL1* (Heuer et al. 2009). This gene confers the enhancement of initial root growth, which enables plants to acquire more phosphorus and increase grain yields in P-deficient soils (Gamuyao et al. 2012).

In this project, we developed a new variety by introgressing the *Pup1* locus into IR64, a major high-yield modern variety in the tropics, in collaboration with the International Rice Research Institute and FOFIFA (National Center for Applied Research on Rural Development). The introgressed breeding lines (*Pup1*-NILs) were transferred to FOFIFA in Madagascar in 2016, and selection and generation advancement in multiple farmer fields was repeated for six consecutive seasons. The selected lines entered variety registration procedures in Madagascar at multiple locations over the three seasons from 2019 to 2021. These procedures were supervised by the variety registration authority of Madagascar, the Service Officiel de Contrôle des Semences et matériel végétal, and involved local farmers. As a result of the production and DHS (Distinctness, Homogeneity, and Stability) tests in 18 sites and organoleptic tests on 400 consumers in 4 villages, one *Pup1*-NIL line was approved for variety release (Fig. 2.6). The new variety was named "FyVary32" after the project's short name "FyVary," which means "Good rice" in the Malagasy language.

FyVary32 had a 12% higher grain yield, a 4-day shorter growth duration on average across 18 sites, and equivalent organoleptic scores relative to X265, one of the major high-yielding varieties in Madagascar. Thereafter, foundation and certified seeds were produced by FOFIFA and seed production farmers, respectively. In the 2022–2023 season, approximately 20 t of certi-

Fig. 2.6 Production test (on the left) and organoleptic test (on the right) for variety registration

Fig. 2.7 FyVary32 grown in a local farmer's field in Madagascar

fied FyVary32 seeds were produced and cultivated by local farmers (Fig. 2.7). This study was the first case that a variety using *Pup1*, the major QTL for phosphorus deficiency tolerance, was officially released and cultivated by local farmers in SSA.

Aboveground morphology is also likely to have a certain function in increasing rice productivity under P deficiency. Rice plants exhibit unique aboveground morphological changes under P deficiency, including reduced tillering, narrow leaves, erect leaves, and retarded leaf elongation rates (Dobermann and Fairhurst 2000; Luquet et al. 2005). The inhibition of tiller bud outgrowth under P deficiency is partly controlled by the strigolactone pathway and is considered an adaptive strategy by rice plants to conserve limited P resources, allowing existing shoots to become productive tillers (Umehara et al. 2010).

However, our studies imply that excessive inhibition of tiller numbers is not always advantageous for rice production under P deficiency (Takai 2024; Tsujimoto et al. 2020; Ishizaki et al. 2023). Ishizaki et al. (2023) controlled the number of tillers by introducing mutations into the coding region of rice *TEOSINTE BRANCHED1* (*OsTB1*), a negative regulator of tillering—using CRISPR/Cas9. We observed that a moderate increase in the number of tillers with the in-frame mutation produced a significantly higher rice grain yield under P deficiency (but not under P sufficiency). The frameshift mutation unlimitedly increased the tiller number but had no advantage in productivity, irrespective of P deficiency, because of many unproductive tillers and sterile grains. Concurrently, Takai et al. (2023) cloned the QTL for *MORE PANICLE 3* (*MP3*) as a natural allele of *OsTB1* from the temperate japonica variety Koshihikari and demonstrated that *MP3* moderately promoted tillering and increased panicle number without negatively affecting panicle size or culm thickness in high-yielding indica cultivars (Fig. 2.8).

We confirmed the effect of *MP3* in increasing panicle number even in P-deficient lowlands in Madagascar (Takai et al. 2021) and then developed a near-isogenic line (X265-*MP3*) by introducing *MP3* to X265, a major high-yielding variety in Madagascar, using backcrossing and marker-assisted selection. Currently, the developed X265-*MP3* is under evaluation for variety registration, using the same procedures as those described above for Madagascar. These results extend the scope of breeding activities to optimize the tillering capacity of rice under nutrient-deficient conditions.

2.5 Future Prospects

The project provided technical solutions for increasing rice yields with little fertilizer input to nutrient-deficient soils in Madagascar. Local farmers are adopting P-dipping techniques and

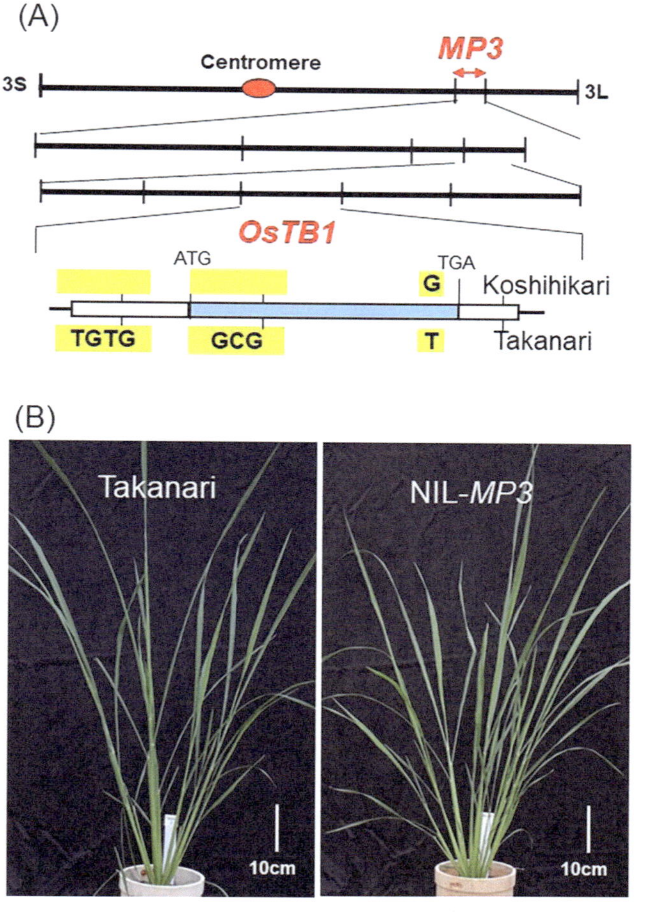

Fig. 2.8 Map-based cloning of *MP3*: sequence differences exist in three locations highlighted in yellow (**a**) and tillering of Takanari, a high-yielding indica cultivar, and the NIL of *MP3* (**b**)

the new variety. Moreover, a concomitant socio-economic survey identified that increased rice yields improve farmers' cash revenues and nutritional status in terms of both energy and micronutrients by accelerating the purchase of nutritious food items, such as vegetables, fruits, and meat/fish, in Madagascar (Nikiema et al. 2023). This economic assessment ensured the validity of technical improvements in rice production to eradicate poverty and hunger in the country. To expand the impact of developed technologies, both the market and extension services should play pivotal roles in ensuring the proper use of technologies and improved access to seeds and fertilizers. Then, a new technical cooperation project funded by JICA started to address these issues and disseminate the P-dipping technique and new variety in collaboration with multiple stakeholders. This transdisciplinary approach will attract further attention to smoothly link research output to beneficiaries and contribute to sustainable agricultural development in SSA.

Acknowledgments I am grateful to all the project members and relevant partners in Japan and in Madagascar who supported our study.

References

Abay KA, Abay MH, Amare M, Berhane G, Aynekulu E (2022) Mismatch between soil nutrient deficiencies and fertilizer applications: implications for yield responses in Ethiopia. Agric Econ 53(2):215–230. https://doi.org/10.1111/agec.12689

Alexander P, Arneth A, Henry R, Maire J, Rabin S, Rounsevell MDA (2023) High energy and fertilizer prices are more damaging than food export curtailment from Ukraine and Russia for food prices, health and the environment. Nat Food 4:84–95. https://doi.org/10.1038/s43016-022-00659-9

Baptista D, Farid M, Fayad D, Kemoe L, Lanci L, Mitra P, Muehlschlegel T, Okou C, Spray J, Tuitoek K, Unsal F (2022) Climate change and chronic food insecurity in sub-Saharan Africa. Departmental paper. DP/2022/016

Batjes NH (2011) ISRIC Report 2011/06: global distribution of soil phosphorus retention potential. Wageningen

Campos P, Borie F, Cornejo P, Lopez-Raez JA, Lopez-Garcia A, Seguel A (2018) Phosphorus acquisition efficiency related to root traits: is mycorrhizal symbiosis a key factor to wheat and barley cropping? Front Plant Sci 9:752. https://doi.org/10.3389/fpls.2018.00752

Chen Z, Balasus N, Lin H, Nesser H, Jacob DJ (2024) African rice cultivation linked to rising methane. Nat Clim Chang 14:148–151. https://doi.org/10.1038/s41558-023-01907-x

Dahlgren RA, Saigusa M, Ugolini FC (2004) The nature, properties and management of volcanic soils. Adv Agron 82:113–182. https://doi.org/10.1016/S0065-2113(03)82003-5

De Vos K, Janssens C, Jacobs L, Campforts B, Boere E, Kozicka M, Havlík P, Folberth C, Balkovic J, Maertens M, Govers G (2023) Rice availability and stability in Africa under future socio-economic development and climatic change. Nat Food 4:518–527. https://doi.org/10.1038/s43016-023-00770-5

Dobermann A, Fairhurst TH (2000) Rice: nutrient disorders & nutrient management. Potash & Phosphate Institute (PPI), Potash & Phosphate Institute of Canada (PPIC) and International Rice Research Institute (IRRI), pp 1–191

FAO, IFAD, UNICEF, WFP, WHO (2024) The State of Food Security and Nutrition in the World 2024 – financing to end hunger, food insecurity and malnutrition in all its forms. Rome. https://doi.org/10.4060/cd1254en

FAOSTAT (2024) Food and agriculture data. https://www.fao.org/faostat/en/#data. Accessed 15 Nov 2024

Gamuyao R, Chin JH, Pariasca-Tanaka J, Pesaresi P, Catausan S, Dalid C, Slamet-Loedin I, Tecson-Mendoza EM, Wissuwa M, Heuer S (2012) The protein kinase Pstol1 from traditional rice confers tolerance of phosphorus deficiency. Nature 488:535–539. https://doi.org/10.1038/nature11346

Heuer S, Lu X, Chin JH, Tanaka JH, Kanamori H, Matsumoto T, De Leon T, Ulat VJ, Ismail AM, Yano M, Wissuwa M (2009) Comparative sequence analyses of the major quantitative trait locus phosphorus uptake 1 (*Pup1*) reveal a complex genetic structure. Plant Biotechnol J 7:456–471. https://doi.org/10.1111/j.1467-7652.2009.00415.x

Ishizaki T, Ueda Y, Takai T, Maruyama K, Tsujimoto Y (2023) In-frame mutation in rice TEOSINTE BRANCHED1 (OsTB1) improves productivity under phosphorus deficiency. Plant Sci 330:11627. https://doi.org/10.1016/j.plantsci.2023.111627

Ismail AM, Singh US, Singh S, Dar MH, Mackill DJ (2013) The contribution of submergence-tolerant (Sub1) rice varieties to food security in flood-prone rainfed lowland areas in Asia. Field Crop Res 152:83–93. https://doi.org/10.1016/j.fcr.2013.01.007

Jayne TS, Mather D, Mason N, Ricker-Gilbert J (2013) How do fertilizer subsidy programs affect total fertilizer use in sub-Saharan Africa? Crowding out, diversion, and benefit/cost assessments. Agric Econ 44(6):687–703. https://doi.org/10.1111/agec.12082

Kihara J, Nziguheba G, Zingore S, Coulibaly A, Esilaba A, Kabambe V, Njoroge S, Palm C, Huising J (2016) Understanding variability in crop response to fertilizer and amendments in sub-Saharan Africa. Agric Ecosyst Environ 229:1–12. https://doi.org/10.1016/j.agee.2016.05.012

Luquet D, Zhang BG, Dingkuhn M, Dexet A, Clément-Vidal A (2005) Phenotypic plasticity of rice seedlings: case of phosphorus deficiency. Plant Prod Sci 8(2):145–151. https://doi.org/10.1626/pps.8.145

Marenya PP, Barret CB (2009) Soil quality and fertilizer use rates among smallholder farmers in western Kenya. Agric Econ 40(5):561–572. https://doi.org/10.1111/j.1574-0862.2009.00398.x

Nikiema RA, Shiratori S, Rafalimanantsoa J, Ozaki R, Sakurai T (2023) How are higher rice yields associated with dietary outcomes of smallholder farm households of Madagascar? Food Sec 15:823–838. https://doi.org/10.1007/s12571-022-01333-5

Nishigaki T, Tsujimoto Y, Rinasoa S, Rakotoson T, Andriamananjara A, Razafimbelo T (2019) Phosphorus uptake of rice plants is affected by phosphorus forms and physicochemical properties of tropical weathered soils. Plant Soil 435:27–38. https://doi.org/10.1007/s11104-018-3869-1

Nishigaki T, Tsujimoto Y, Rakotoson T, Rabenarivo M, Andriamananjara A, Asai H, Andrianary HB, Rakotonindrinan H, Razafimbelo T (2021) Soil phosphorus retention can predict responses of phosphorus uptake and yield of rice plants to P fertilizer application in flooded weathered soils in the central highlands of Madagascar. Geoderma 402:115326. https://doi.org/10.1016/j.geoderma.2021.115326

Nishigaki T, Tsujimoto Y, Rakotonindrina H, Andriamananjara A (2023) Estimation of P retention capacity by the water content of soil kept with a saturated NaCl solution in a desiccator. Soil Sci Plant Nutr 69(5–6):337–345. https://doi.org/10.1080/00380768.2023.2245420

Nziguheba G, Zingore S, Kihara J, Merckx R, Njoroge S, Otinga A, Vandamme E, Vanlauwe B (2016) Phosphorus in smallholder farming systems of sub-Saharan Africa: implications for agricultural intensification. Nutr Cycl Agroecosyst 104:321–340. https://doi.org/10.1007/s10705-015-9729-y

Oo AZ, Tsujimoto Y (2022) Localized phosphorus application via P-dipping is more effective for improving initial rice growth in lower temperature conditions. Plant Prod Sci 26(1):28–35. https://doi.org/10.1080/1343943X.2022.2160363

Oo AZ, Tsujimoto Y, Rakotoarisoa NM (2020a) Optimizing the phosphorus concentration and duration of seedling dipping in soil slurry for accelerating the initial growth of transplanted rice. Agronomy 10(2):240. https://doi.org/10.3390/agronomy10020240

Oo AZ, Tsujimoto Y, Rakotoarisoa NM, Kawamura K, Nishigaki T (2020b) P-dipping of rice seedlings increases applied P use efficiency in high P-fixing soils. Sci Rep 10:11919. https://doi.org/10.1038/s41598-020-68977-1

Oo AZ, Tsujimoto Y, Rakotoarisoa NM, Andrianary BH (2023) Localized phosphorus application via P-dipping doubles applied P use efficiency and avoids weather-induced stresses for rice production on P-deficient lowlands. Eur J Agron 149:126901. https://doi.org/10.1016/j.eja.2023.126901

Ozaki R, Sakurai T (2021) Profitability of chemical fertilizer application: comparison of lowland and upland rice cultivation in Madagascar. Jpn J Agric Econ 23:119–124. https://www.jstage.jst.go.jp/article/jjae/23/0/23_119/_article/-char/en

Ozaki R, Tsujimoto Y, Andriamananjara A, Rakotonindrina H, Sakurai T (2024) Optimizing fertilizer use by providing soil quality information: experimental evidence from Madagascar. Agric Food Secur 13:45. https://doi.org/10.1186/s40066-024-00500-5

Rakotoarisoa NM, Tsujimoto Y, Oo AZ (2020) Dipping rice seedlings in P-enriched slurry increases grain yield and shortens days to heading on P-deficient lowlands in the central highlands of Madagascar. Field Crop Res 254:107806

Rakotoarisoa NM, Tsujimoto Y, Oo AZ, Tashiro T, Kano-Nakata M, Ehara H (2023) Dipping vigorous seedling roots in phosphorus-enriched slurry at transplanting efficiently increases lowland rice yields. Crop Environ 2(4):202–208. https://doi.org/10.1016/j.crope.2023.10.002

Rakotoson T, Tsujimoto Y, Nishigaki T (2022) Phosphorus management strategies to increase lowland rice yields in sub-Saharan Africa: a review. Field Crop Res 275(2022):108370. https://doi.org/10.1016/j.fcr.2021.108370

Ricker-Gilbert J, Jayne TS (2012) Do fertilizer subsidies boost staple crop production and reduce poverty across the distribution of smallholders in Africa? Quantile regression results from Malawi. Selected paper for the triennial meeting of the International Association of Agricultural Economists: Foz Do Iguacu, Brazil. https://doi.org/10.22004/ag.econ.126742

Saito K, Vandamme E, Johnson J-M, Tanaka A, Senthilkumar K, Dieng I, Akapko C, Gbaguidi F, Segda Z, Bassoro I, Lamare D, Gbakatchetche H, Abera BB, Jaiteh F, Bam RK, Dogbe W, Sékou K, Rabeson R, Kamissoko N, Mossi IM, Wopereis MCS (2019) Yield-limiting macronutrients for rice in sub-Saharan Africa. Geoderma 338:546–554. https://doi.org/10.1016/j.geoderma.2018.11.036

Saito K, Senthilkumar K, Dossou-Yovo ER, Ali I, Johnson J-M, Mujawamariya G, Rodenburg J (2023) Status quo and challenges of rice production in sub-Saharan Africa. Plant Prod Sci 26(3):320–333. https://doi.org/10.1080/1343943X.2023.2241712

Sanchez PA (2015) En route to plentiful food production in Africa. Nat Plants:14014. https://doi.org/10.1038/nplants.2014.14

Takai T (2024) Potential of rice tillering for sustainable food production. J Exp Bot 75(3):708–720. https://doi.org/10.1093/jxb/erad422

Takai T, Sakata M, Rakotoarisoa NM, Razafinarivo NT, Nishigaki T, Asai H, Ishizaki T, Tsujimoto Y (2021) Effects of quantitative trait locus MP3 on the number of panicles and rice productivity in nutrient-poor soils of Madagascar. Crop Sci 61:519–528. https://doi.org/10.1002/csc2.20344

Takai T, Taniguchi Y, Takahashi M et al (2023) *MORE PANICLES 3*, a natural allele of *OsTB1/FC1*, impacts rice yield in paddy fields at elevated CO_2 levels. Plant J 114:729–742. https://doi.org/10.1111/tpj.16255

Tsujimoto Y, Inusah B, Katsura K, Fuseini A, Dogbe W, Zakaria AI, Fujihara Y, Oda M, Sakagami J-I (2017) The effect of sulfur fertilization on rice yields and nitrogen use efficiency in a floodplain ecosystem of northern Ghana. Field Crop Res 211:155–164. https://doi.org/10.1016/j.fcr.2017.06.030

Tsujimoto Y, Rakotoson T, Tanaka A, Saito K (2019) Challenges and opportunities for improving N use efficiency for rice production in sub-Saharian Africa. Plant Prod Sci 22(4):413–427. https://doi.org/10.1080/1343943X.2019.1617638

Tsujimoto Y, Sakata M, Raharinivo V, Tanaka JP, Takai T (2020) AZ-97 (*Oryza sativa* ssp. *Indica*) exhibits superior biomass production by maintaining the tiller numbers, leaf width, and leaf elongation rate under phosphorus deficiency. Plant Prod Sci 24(1):41–51. https://doi.org/10.1080/1343943X.2020.1808026

Tsujimoto Y, Tanaka A, Rakotoson T (2021) Sequential micro-dose fertilization strategies for rice production: improved fertilizer use efficiencies and yields on P-deficient lowlands in the tropical highlands. Euro J Agron 131:126381. https://doi.org/10.1016/j.eja.2021.126381

Umehara M, Hanada A, Magome H, Takeda-Kamiya N, Yamaguchi S (2010) Contribution of strigolactones to the inhibition of tiller bud outgrowth under phosphate deficiency in rice. Plant Cell Physiol 51(7):1118–1126. https://doi.org/10.1093/pcp/pcq084

Van Campenhout B (2021) The role of information in agricultural technology adoption: experimental evidence from rice farmers in Uganda. Econ Dev Cult Change 69(3):1239–1272. https://doi.org/10.1086/703868

van Oort PAJ (2018) Mapping abiotic stresses for rice in Africa: drought, cold, iron toxicity, salinity and sodicity. Field Crop Res 219:55–75. https://doi.org/10.1016/j.fcr.2018.01.016

Wissuwa M (2005) Combining a modelling with a genetic approach in establishing associations between genetic and physiological effects in relation to phosphorus uptake. Plant Soil 269:57–68. https://doi.org/10.1007/s11104-004-2026-1

Wissuwa M, Ae N (2001) Genotypic variation for tolerance to phosphorus deficiency in rice and the potential for its exploitation in rice improvement. Plant Breed 120:43–48. https://doi.org/10.1046/j.1439-0523.2001.00561.x

Wissuwa M, Yano M, Ae N (1998) Mapping of QTLs for phosphorus-deficiency tolerance in rice (*Oryza sativa* L.). Theor Appl Genet 97:777–783. https://doi.org/10.1007/s001220050955

World Bank (2023) Food security update. World Bank Publications, Washington, DC. https://thedocs.worldbank.org/en/doc/40ebbf38f5a6b68bfc11e5273e1405d4-0090012022/related/Food-Security-Update-LXXXI-March-23-2023.pdf

Zhang X, Davidson EA, Mauzerall DL, Searchinger TD, Dumas P, Shen Y (2015) Managing nitrogen for sustainable development. Nature 528:51–59. https://doi.org/10.1038/nature15743

Zingore S, Tittonell P, Corbeels M, van Wijk MT, Giller KE (2011) Managing soil fertility diversity to enhance resource use efficiencies in smallholder farming systems: a case from Murewa District Zimbabwe. Nutr Cycl Agroecosyst 90(1):87–103. https://doi.org/10.1007/s10705-010-9414-0

Zou T, Zhang X, Davidson EA (2022) Global trends of cropland phosphorus use and sustainability challenges. Nature 611:81–87. https://doi.org/10.1038/s41586-022-05220-z

3 Rice Research for Tailor-Made Breeding and Cultivation Technology Development in Kenya

Daigo Makihara, Emily Gichuhi, Daniel Menge, Ruth Musila, Kazuyuki Doi, Mayumi Kikuta, John Kimani, Yoshiaki Inukai, Hiroaki Samejima, Rena Tomita, and Akira Yamauchi

Abstract

Rice production in Kenya faces numerous challenges, including drought, cold damage, salinity, low-fertility soils, and blast disease. To address these biotic and abiotic stresses and boost rice production, the Rice Research for Tailor-Made Breeding and Cultivation Technology Development project was conducted from 2013 to 2018. This project aimed to establish a research base for breeding and cultivation technology development in Kenya. The objectives were to develop a rice breeding and evaluation system, characterize rice varieties to identify beneficial traits, develop breeding lines using beneficial genes/QTLs, identify challenges and technical solutions to enhance rice productivity, and identify the optimal conditions for effective gene/QTL expression through genotype × environment × management interaction analysis. Key outcomes included upgrading the research environment at KALRO Mwea, establishing evaluation systems for various traits,

D. Makihara · Y. Inukai · R. Tomita
International Center for Research and Education in Agriculture, Nagoya University, Nagoya, Aichi, Japan
e-mail: makihara@agr.nagoya-u.ac.jp; inukaiy@agr.nagoya-u.ac.jp; tomita@agr.nagoya-u.ac.jp

E. Gichuhi · R. Musila
Kenya Agricultural and Livestock Research Organization, Industrial Crops Research Institute-Mwea, Kerugoya, Kirinyaga County, Kenya

D. Menge
International Rice Research Institute, Africa Regional Office-Nairobi, Nairobi, Kenya

K. Doi
Graduate School of Bioagricultural Sciences, Nagoya University, Nagoya, Aichi, Japan
e-mail: kdoi@agr.nagoya-u.ac.jp

M. Kikuta
Faculty of Agriculture, Tokyo University of Agriculture, Atsugi, Kanagawa, Japan
e-mail: mk208750@nodai.ac.jp

J. Kimani
Kenya Agricultural and Livestock Research Organization, Food Crops Research Institute-Mtwapa, Mtwapa, Kilifi County, Kenya

H. Samejima
Hokkaido Agricultural Research Center, National Agriculture and Food Research Organization, Crop Science, Sapporo, Hokkaido, Japan
e-mail: samejimah904@naro.affrc.go.jp

A. Yamauchi (✉)
Asian Satellite Campuses Institute, Nagoya University, Nagoya, Aichi, Japan
e-mail: ayama@agr.nagoya-u.ac.jp

T. Nagamine et al. (eds.), *Plant Genetic Resources (PGR) for Sustainable Crop Production*, Sustainable Development Goals Series, https://doi.org/10.1007/978-981-96-7117-5_3

identifying useful breeding materials, developing promising breeding lines, and demonstrating technical improvement measures. Continued efforts after the project have led to the registration of new rice varieties in Kenya and the strengthening of the rice seed production system. This project also contributed to network building and human resource development. Future prospects include continuing variety improvement and cultivation technology development, incorporating cutting-edge scientific technologies, enhancing the research capacity at KALRO Mwea, and developing it as a hub for rice research talent in East Africa.

Keywords

DNA marker-assisted selection · Biotic and abiotic stress · Crop trait evaluation · Seed production system · G × E × M interaction · Variety registration

3.1 Background

In many sub-Saharan African countries, domestic rice production has not kept pace with increasing demand, making rice production a critical issue for food security in these regions. In Kenya, rice consumption surged after 1990, and annual imports have consistently exceeded 600,000 t since 2018. Consequently, increasing domestic production is an urgent challenge.

Rice cultivation in Kenya was hindered by numerous obstacles, including drought, cold damage in high-altitude regions, salinity, low-fertility soils, blast disease, and rice yellow mottle virus (RYMV). To address these biotic and abiotic stresses and boost rice production, it is essential to develop stress-resistant varieties.

Recent advancements in genetics, breeding, and crop sciences have identified various traits that are beneficial for rice production and associated genes and quantitative trait loci (QTLs). These developments have made it technically feasible to improve rice varieties by the tailor-made introduction of such traits. Specifically, by utilizing DNA markers, it is possible to efficiently select individuals possessing the desired genes/QTLs from lines developed through crossing and create new crop varieties.

However, stress tolerance and productivity in real cultivation settings are not solely determined by the genetic factors of a variety. They are also influenced by cultivation environment and management practices. To develop varieties and cultivation technologies that are adaptable to local conditions, it is essential to identify traits and their associated genes/QTLs that function effectively under local cultivation environments and to develop varieties that incorporate these traits. Simultaneously, it is crucial to establish cultivation techniques that maximize the potential of these varieties.

To achieve this, cultivation experiments should be conducted under local environmental conditions. This includes analyzing how environmental conditions and cultivation practices affect the functional expression of genes/QTLs. Therefore, the primary goal of this project was to establish a research base for breeding and cultivation technology development in Kenya.

This SATREPS project was conducted from May 2013 to May 2018, with Nagoya University as the leading research institution on the Japanese side and the Kenya Agricultural and Livestock Research Organization (KALRO) representing Kenya. The collaborative research institutions included Okayama University, Shimane University, and Yamagata University. This research project aimed to achieve the following objectives:

1. Development of a rice breeding and evaluation system

   Establishing a foundation for crossbreeding, DNA marker-assisted selection, and population breeding in Kenya. This involved developing evaluation fields, selecting reference varieties, and setting evaluation criteria to evaluate key traits, including drought resistance, cold tolerance, salt tolerance, adaptation to low-fertility soils, and resistance to blast disease.

2. Characterization of rice varieties and identification of beneficial traits

   Characterizing the existing Kenyan rice varieties and global genetic resources using the evaluation system. This includes detecting QTLs linked to beneficial traits, such as yield improvement or stress tolerance, and validating their use in breeding programs.
3. Development of breeding lines for Kenya using beneficial genes/QTLs

   Introducing beneficial genes/QTLs into locally adapted rice varieties through molecular breeding techniques such as DNA marker-assisted selection, genomic selection with SNP information, and gene editing to develop intermediate parental lines suited to Kenya's agricultural needs.
4. Identification of challenges and technical solutions to enhance rice productivity

   Identifying challenges to improve rice productivity through field surveys on water and soil conditions, along with current cultivation practices in the Mwea Irrigation Scheme, Kenya's largest rice-growing region, and developing effective measures to address these challenges.
5. Analysis of genotype (G) × environment (E) × management (M) interactions

   Identifying optimal conditions for effective gene/QTL expression to develop cultivation techniques that maximize the potential of lines with beneficial genes/QTLs through analysis of genotype (G), environment (E), and management practices (M).

A more detailed description of the issues and potential of rice cultivation in sub-Saharan Africa outlined above, the significance of conducting joint research in Kenya, the objectives of this SATREPS project, and the approaches to addressing the challenges, as well as the progress of the project as of 2017 can be found in Makihara et al. (2018).

3.2 Key Outcomes of the Project

3.2.1 Establishment of a Rice Research Base in Kenya

Kenya, situated on the eastern coast of Africa and straddling the equator, has rice-growing areas ranging from coastal lowlands to highlands, at altitudes of approximately 1,400 m. This diversity of cultivation environments supports various rice farming systems, including traditional rainfed lowland farming, modern irrigated paddies, and upland rice cultivation. These diverse farming systems present a wide range of production challenges. Kenya experiences most of the biotic and abiotic stresses that affect rice production in sub-Saharan Africa. Kenya's diverse rice-growing conditions make it an ideal location for establishing a research base to address the rice production challenges in sub-Saharan Africa.

At the start of the project, the research environment at KALRO Mwea was underdeveloped, with no modern experimental facilities and fields unsuitable for precise cultivation trials. To address these limitations, we began upgrading the research environment at KALRO Mwea.

The construction of laboratories for rice research was funded by the KR Counterpart Fund under Japan's Food Aid program. Completed in 2016, the facility consists of a molecular genetics laboratory, general laboratory, yield survey room, cold room, storage rooms, meeting rooms, and researcher offices (Fig. 3.1). Through the SATREPS project, various experimental equipment and facilities were installed to support cultivation trials, genetic analyses, chemical analyses for soil and plant samples, physiological measurements, morphological observations, and crossbreeding activities. Furthermore, systems for evaluating traits such as cold tolerance, drought resistance, blast disease resistance, adaptation to low-fertility soils, and salt tolerance were established.

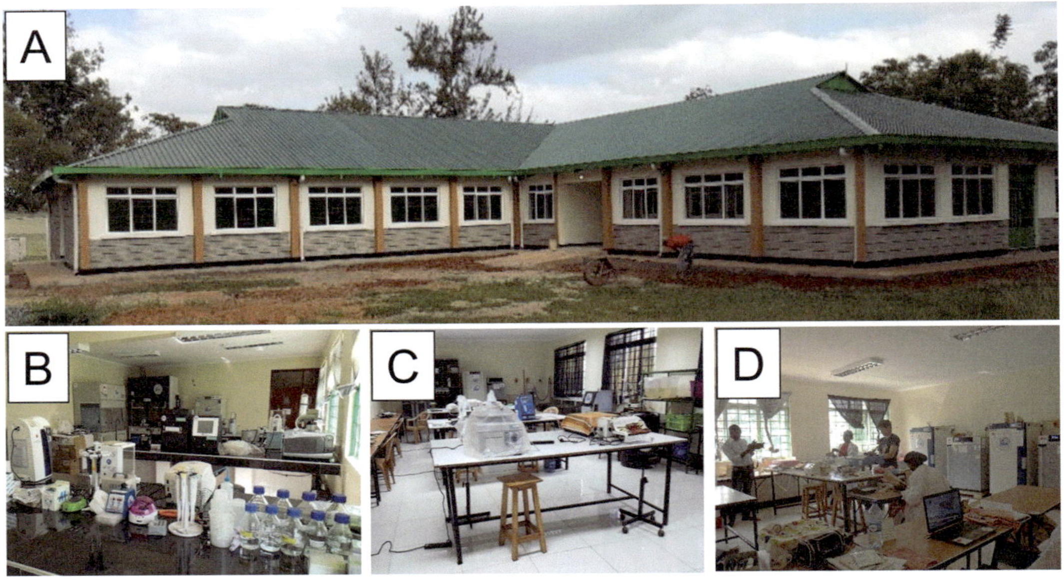

Fig. 3.1 The building for rice research (**a**), constructed at Kenya Agricultural and Livestock Research Organization Mwea in 2016 using the KR Counterpart Fund under Japan's Food Aid program, and its molecular genetics laboratory (**b**), general laboratory (**c**), and plant processing laboratory (**d**)

For high-precision cultivation trials, essential infrastructure, including seedling facilities, greenhouses, meteorological equipment, and irrigation systems, was developed at KALRO Mwea. Concrete-framed paddy fields with irrigation systems and bird-proof netting were developed for multi-location trials at KALRO Mwea, KALRO Kibos, and KALRO Mtwapa through a follow-up project entitled *Improvement of rice production by tailor-made breeding and cultivation technology development in Africa*, supported by the Accelerating Social Implementation for SDGs Achievement (aXis) program of the Japan Science and Technology Agency (JST). Agricultural machinery, including tractors, power tillers, a combine harvester, a machinery trailer, reapers, a rice milling machine, a rice mill house, grain dryers, drying floors, and a seed dresser, was also introduced under the SATREPS and aXis projects (Fig. 3.2).

Even after the completion of the SATREPS project, we have been leveraging the rice research base we developed to expand our collaborative research activities, addressing challenges such as low productivity, stress tolerance, and sustainability in rice production in Kenya and beyond.

3.2.2 Development of Rice Breeding and Variety Evaluation System in Kenya

As the core of breeding activities, the candidate breeding lines produced in the project, in addition to the representative varieties and the donors of useful alleles, have been maintained and grown continuously for the use of crossing and evaluation. The system for producing genetic/breeding materials requires crossing and selection using DNA markers. The method of crossing (pollination) needed to be adjusted to the environment of Kenya. Two methods for emasculation, the vacuum method and the hot water method, have been employed by the introduction of necessary instruments. However, flower organs were damaged by very low humidity, leading to a low seed set in pollinated panicles. The use of plastic bags after pollination to prevent drying mitigated this problem. The selection for pyramiding lines was enabled by DNA marker laboratory implemented in the project. Currently leaf sampling and DNA extraction can be done routinely. The leaf samples for DNA extraction were oven-dried, and the

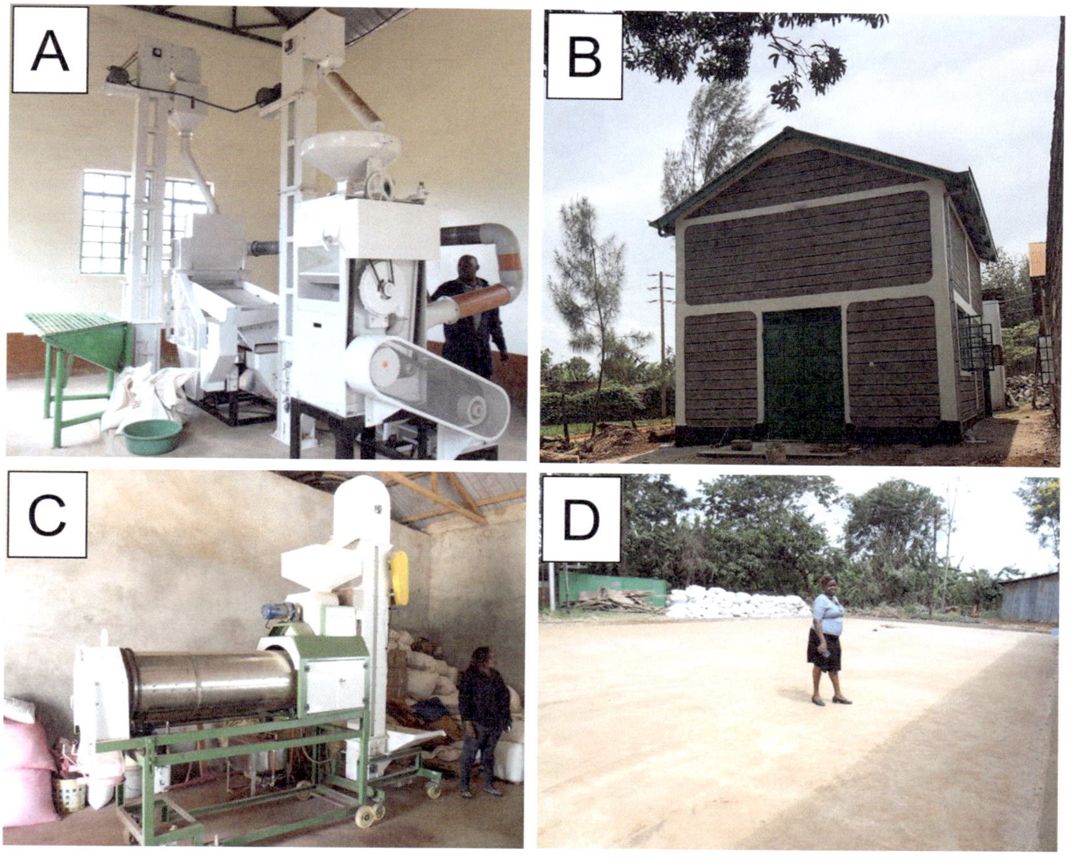

Fig. 3.2 Facilities for seed production installed at Kenya Agricultural and Livestock Research Organization Mwea in 2021: rice milling machine (**a**), rice mill house (**b**), seed dresser (**c**), and drying floor (**d**)

storage method has been simplified to one that does not require a freezer. Significant progress has been made in establishing the DNA marker laboratory, and efforts are ongoing to further enhance its operation, particularly in ensuring a steady supply of reagents and consumables, which remain challenging to procure in Kenya. Furthermore, notable advancements are being made in the application of population breeding. A trial focusing on cold-tolerant lines derived from a cross between Basmati 370 and Lijiangxintuanheigu is underway, showcasing the promising potential to expand the use of population breeding method alongside the marker-assisted backcrossing approach.

The breeding for cold tolerance mentioned above has been conducted using the cold tolerance evaluation system we developed in Mwea. This system is based on a multiple-sown field evaluation method, which was established to effectively evaluate the cold tolerance in rice under tropical highland conditions in Mwea. New Rice for Africa (NERICA) and their parent varieties were cultivated 10 times to expose them to a shift in temperature during the 20-day periods before heading (critical period), measuring the average daily mean temperature during the critical period (Tmean) for each variety, and the filled grain ratio, which was used as an indicator of cold tolerance. Tmean of each variety was lower in six out of ten cultivations (20.1–21.2 °C) than in the remaining four cultivations (21.3–23.3 °C). There were significant varietal differences in the filled grain ratio only in the six low-temperature cultivations. The results of cold tolerance evaluation in this study were generally in line with the previous results under temperature-controlled environments, showing the practicality of multiple-sown field cold tolerance evaluation in Mwea. The multiple sowing was necessary to overlap the critical period in each variety and low-temperature period and to increase the number of varieties which experience virtually the same cold stress level. Additionally, the set of NERICAs and their parent varieties, which recorded different levels of cold tolerance and whose average number of days to heading in all experiments were 66–78, would be suitable as standard check varieties in cold tolerance evaluation of rice varieties in this region. These results have been reported by Samejima et al. (2020).

As an evaluation system for blast resistance, we established a field-level evaluation method that can be maintained even in Mwea, where research facilities are limited. Over 4 years of cultivation trials, we demonstrated that stable blast disease development can be achieved under field conditions by cultivating highly susceptible varieties as spreader plants without pesticides, placing infected rice straw from the previous year around the base of rice plants, and maintaining high humidity through daily sprinkler irrigation during the short rainy season (October to December). In this evaluation field, we can investigate the blast resistance of target varieties or lines along with a new set of international differential varieties (DVs), composed of monogenic lines carrying 23 different resistance genes (Tsunematsu et al. 2000). Based on the reaction patterns of DVs to blast isolates present in the field, we can estimate the pathogenicity of the isolates and the resistance genotypes of the test materials (Fukuta et al. 2019b).

In addition, we developed several specialized fields to assess various traits, including a sloped field for evaluating deep rooting associated with drought resistance in upland rice; concrete beds filled with four distinct soil types with significantly different physical properties, such as permeability and penetration resistance, to evaluate the expression of root system traits related to drought resistance across different soil environments in Kenya; concrete beds within a greenhouse filled with saline-affected soils to evaluate salt tolerance; unfertilized low-fertility paddies maintained without fertilization since the beginning of the SATREPS project to assess performance under nutrient-deficient conditions; low-phosphorus upland fields for evaluating phosphorus deficiency tolerance; and paddy fields designed for comparative experiments using two distinct soil types (Vertisols and Nitisols). Additionally, we established experimental paddy fields to study the long-term effects of varying application rates of nitrogen, phosphorus, and potassium, as well as the effects of no fertilization and compost application.

3.2.3 Characterization of Rice Varieties and Identification of Beneficial Traits

To identify useful breeding materials for variety improvement in Kenya, we evaluated various characteristics of existing rice varieties, including Kenya's local landraces, using the trait evaluation system developed under this project. We provide detailed examples of these evaluations below.

The upland rice varieties NERICA 1 and NERICA 4 are well known for their high drought resistance and are widely cultivated across sub-Saharan Africa, including Kenya. We evaluated their root traits related to drought resistance under different soil conditions: soils that allow deep root growth and soils with an impediment for deep root development. The results revealed distinct differences in root characteristics between the two varieties. Specifically, NERICA 1 demonstrated the ability to efficiently absorb water from the soil surface, such as rainfall, by developing a higher number of lateral roots under moderate drought conditions. In contrast, NERICA 4 showed greater ability to utilize water from deeper soil layers by growing deeper roots (Menge et al. 2016). In drought resistance breeding, it is important to introduce appropriate root traits tailored to the target soil environment, and both varieties can serve as valuable breeding materials for this purpose.

Using the cold tolerance evaluation system developed in Mwea, we assessed more than 100 rice varieties and lines. The results revealed that the major Kenyan varieties—Basmati 370, Basmati 217, ITA 310, BW196, and IR2793-80-1—were all cold-susceptible. Furthermore, we demonstrated that the NERICA varieties show a wide variation in cold tolerance, ranging from very tolerant to very susceptible, and identified a QTL for cold tolerance from WAB56-104, a parental variety of NERICA.

In the blast resistance evaluation field established at KALRO Mwea, we estimated the blast races in the region by evaluating disease severity on differential varieties (DVs) targeting 23 blast resistance genes and a susceptible check variety, Lijiangxintuanheigu. Moreover, we investigated the pathogenicity of 99 isolates of rice blast (*Pyricularia oryzae* Cavara) collected from 5 regions in Kenya (Kirinyaga County, Embu County, Kisumu County, Tana River County, and Mombasa County) based on the reaction patterns of the 23 DVs and Lijiangxintuanheigu in blast inoculation tests. The results revealed that blast isolates virulent to resistance genes *Pib*, *Pia*, *Pii*, *Pi3*, *Pi5*(t), *Pik-s*, *Pik-m*, *Pi1*, *Pik-h*, *Pik*, *Pik-p*, *Pi7*(t), *Pi19*(t), and *Pi20*(t) were the predominant types in many rice-growing areas of Kenya (Fukuta et al. 2019b). However, in the southern coastal areas of the Indian Ocean, where diverse rice varieties are cultivated in small rainfed paddy fields, there was no dominant type observed.

Furthermore, we investigated the genetic variation of 47 rice accessions collected from Kenya using polymorphism data from simple sequence repeat (SSR) markers and classified these accessions based on their reaction patterns to standard differential blast isolates (SDBIs). The major cultivated varieties in Kenya, Basmati 370, Basmati 217, and ITA 310, were classified into a highly resistant cluster group. However, they were found to be susceptible to some of the SDBIs from Kenya (Fukuta et al. 2019a).

Under the conditions of the blast resistance evaluation field at KALRO Mwea, it was found that the single introgression of the blast resistance gene *pi21*, which has been reported to improve blast resistance in various varieties (Angeles-Shim et al. 2020), into Basmati 370 did not enhance resistance. This suggests that pyramiding it with other resistance genes is necessary.

3.2.4 Development of Breeding Lines for Kenya Using Beneficial Genes/QTLs

The effects of introgressing beneficial QTLs were evaluated, which resulted in the successful production of intermediate parental lines. These included 12 lines expected to enhance rice yield, 3 lines for blast resistance, 4 or more lines for low-fertility soil adaptation, 1 line each for cold tolerance and cold injury avoidance, and 1 line for drought avoidance. In addition, a breeding program was developed, which outlines the process from these intermediate parental lines to variety development and farmer dissemination. The following are examples of these development cases.

To improve yield, we focused on developing lines that incorporate genes such as *Gn1a* and *WFP*, which are associated with an increase in the number of grains per panicle. The alleles were introgressed into various NERICA varieties by using marker-assisted backcrossing (MABC) (Reyes et al. 2021). The introgressed chromosome regions were detected in detail by using genotyping by sequencing (GBS). The phenotypic effects of the introgressions were confirmed at IRRI (Reyes et al. 2021). These materials were evaluated at Mwea, Kenya, for the actual yield, and a NERICA 1-based line (NERICA 1 + *Gn1a* + *WFP*) showed greater performance than the recurrent parent (Kikuta et al. 2023). On the other hand, introgression of *WFP* into IR-BB60, a breeding line with IR24 genetic background, did not result in an enhanced yield because of poor grain filling (Yamada et al. 2020).

We are conducting further research on breeding and cultivation management to fill the gap between the sink (panicle) size and actual yield.

Cold tolerance at the booting stage based on spikelet fertility under cold stress was evaluated using the F_2 population derived from a cross between temperate japonica, Hananomai, and tropical japonica, WAB56-104. Two QTLs for cold tolerance were detected on chromosomes 8 and 10, with enhanced effects on the trait originating from the Hananomai and WAB56-104 alleles, respectively. Selected backcrossed lines having homozygous alleles for both functional cold tolerance QTLs exhibited higher spikelet fertility under both cold-water irrigation in Japan and natural weather conditions in the Kenyan highland. Thus, the identified QTLs will be useful for the development of cold-tolerant varieties for production in high-altitude areas (Wainaina et al. 2018).

Enhanced root development is essential for crop adaptability and productivity under water-deficit conditions. We identified *outstanding rooting1* (*our1*) rice mutant and evaluated its growth performance under alternate wetting and drying (AWD) water-saving irrigation in a Kenyan field. Compared to wild-type genotypes, genotypes harboring *our1* mutation gene possessed a well-developed root system and showed better shoot growth and yield, especially through increased water uptake during rewatered conditions between drought periods in AWD. This finding indicates that the *our1* mutation gene can serve as a novel breeding material to mitigate transient drought stress on yield under AWD (Hasegawa et al. 2022).

Oryza longistaminata Chromosome Segments Introduced Lines (LCSILs) are among the promising lines developed during the SATREPS project. The LCSILs carried chromosome segments from the wild rice genome, *O. longistaminata*, within a Kernel Basmati genetic background. These lines were evaluated for various biotic and abiotic stress tolerances and agronomic performance in Kenya. Evaluation of blast resistance in LCSILs using the system developed at KALRO Mwea and the investigation of genetic variations in resistance genes within LCSILs revealed that LCSIL6, LCSIL12, LCSIL27, LCSIL28, and LCSIL40 possess either the *Pish* or *Piz-t* gene, or both, and exhibited resistance to blast disease races in Kenya (Tomita et al. 2023). These lines serve as valuable sources of key resistance genes against blast disease and are therefore currently being used as breeding lines in the rice breeding program at KALRO Mwea to improve rice varieties, including Basmati 370. Furthermore, evaluation of dry matter productivity and agronomic performance under different soil fertility conditions led to the selection of eight promising lines (LCSIL2, LCSIL4, LCSIL23, LCSIL24, LCSIL38, LCSIL43, LCSIL44, and LCSIL45) for adaptation to low-fertility soils. Research on the selection of lines adapted to low-fertility soils is ongoing.

3.2.5 Identification of Challenges and Technical Solutions to Enhance Rice Productivity

At the start of the project, we conducted irrigation water quality surveys, soil analyses, yield surveys in farmers' fields, and interviews with farmers to identify challenges in rice production in Kenya. We also developed and demonstrated technical improvement measures to address these challenges. Below are some examples of our findings and their contributions.

In Mwea, we found that while irrigation water was adequately supplied in the upstream areas of irrigation canals, interrupted water supply in downstream areas caused yield declines. This issue was attributed to the inefficient distribution of irrigation water and the limited adoption of water-saving technologies, such as proper puddling to reduce vertical water infiltration and efficient irrigation methods. Our project demonstrated that intermittent irrigation could reduce water use and irrigation frequency by up to 50% compared with continuous flooding. Furthermore, appropriate puddling reduced the daily percolation depth to approximately one-tenth of that observed under conventional land preparation methods.

Our investigations in Mwea revealed that higher salinity levels in paddy fields, as indicated by high electrical conductivity (EC), were associated with lower rice yields. Soil analysis data from both the paddy fields within the Mwea Irrigation Scheme and the surrounding outgrower paddy fields showed that salts, particularly calcium (Ca) and magnesium (Mg), accumulated in the downstream paddy fields. This led to a reduced ratio of K to Ca and Mg, resulting in low K availability for rice in many fields. The accumulation of salts and the low availability of K were likely due to practices such as continuous flow irrigation and field-to-field irrigation (where water is channeled from one rice paddy to another without a proper canal), which allowed fertilizer components and salts dissolved in the soil to flow downstream. Additionally, farmers typically applied only nitrogen and phosphorus fertilizers, with little to no application of K.

To address the issue of low K availability in soils, we conducted cultivation trials to assess the effects of K fertilization. These trials demonstrated that increasing K application improved rice yield. This information was widely shared during seminars held in Mwea, which were attended by local rice stakeholders including members of the JICA technical cooperation project on rice (Rice-Based and Market-oriented Agriculture Promotion Project: RiceMAPP). Furthermore, we developed a manual to demonstrate improved cultivation techniques, which provided protocols to validate and demonstrate these practices in farmers' fields.

Information on the seasonal variation in grain yield and climate conditions is necessary to determine the optimal rice cropping calendar in Mwea. We conducted a series of 58 experiments using the popular rice variety Basmati 370 under a secured water supply and adequate blast control, with sowing intervals of 15 days. The results show that the current cropping calendar (July–December or August–January) is suitable for single rice cropping. Double cropping is currently unfeasible due to insufficient water for irrigation. If water shortages are resolved, the introduction of double cropping could increase annual rice production by more than 100%, as the grain yield from additional cultivation (starting in January or February) was comparable to the yield in the current cropping calendar. Sowing Basmati 370 between March and June is not recommended because of suboptimal yield resulting from low temperatures and solar radiation from May to September. These findings have been reported by Samejima et al. (2020).

Nitrogen fertilizer applications above 75 kg ha^{-1}, the current standard N fertilizer amount in Mwea, were ineffective for increasing the grain yield of Basmati 370 and several other high-yielding varieties grown in Kenya and other countries. Shoot dry weight reached one of the highest recorded values globally (20 t ha^{-1} or more) under increased N conditions (>75 kg ha^{-1}) in several cases. However, the increase in biomass did not translate into higher grain yield due to a decreased filled grain ratio under high-N conditions. In contrast, the grain yield of NERICA 1 increased with increasing N fertilizer amounts because this variety minimized the reduction in the filled grain ratio caused by high-N levels. The reasons for varietal differences in the response of the filled grain ratio to high-N conditions remain unclear. Introducing varieties without considering their response to high-N fertilization could hinder efforts to increase rice production in Mwea. These findings were reported by Njinju et al. (2018) and Samejima et al. (2021).

3.2.6 Analysis of the G × E × M Interactions

A notable achievement in the G × E × M interactions in this project involves the distinct root traits associated with drought resistance in NERICA 1 and NERICA 4 and their interactions with fertilization and field conditions. We conducted a series of experiments to investigate the effects of fertilizer management on the expression of different root traits for drought resistance in NERICA 1 and NERICA 4 under different soil conditions: soils that allow deep root growth and soils that impede deep root development.

In the first set of experiments, the effects of nitrogen (N) application on the root and shoot growth of upland NERICA varieties were assessed under field conditions that restricted deep root growth. Under drought conditions, NERICA 1 exhibited a significantly higher shoot dry weight, total root length, and lateral root length than NERICA 4 at adequate N levels (Menge et al. 2019). However, no significant differences were observed between the two NERICA varieties in dry matter production under N deficiency. Under continuous drought, the benefits of N application in alleviating soil moisture stress were observed under moderate drought, but not under severe drought conditions. Under soil moisture fluctuations, NERICA 1 exhibited a greater ability to develop lateral roots during soil drying when sufficient N was applied (Menge et al. 2020). This root growth occurred mainly between the maximum tillering and heading stages. Enhanced root growth under adequate N fertilization contributed to improved shoot growth as water availability increased, particularly during these growth stages.

In the second set of experiments, the effects of phosphorus (P) fertilization on deep root growth and dry matter production in upland rice under constant drought were evaluated using 65-cm-high polyvinyl chloride (PVC) pots and half-split PVC tubes with transparent acrylic plates to observe root growth. The results from the two setups showed that NERICA 4 had greater dry matter production under drought conditions than NERICA 1 at adequate P levels. Under drought, NERICA 4 exhibited greater root length density in the 40–60 cm soil layer when sufficient P fertilization was applied. Moreover, the root elongation rate of the deepest nodal root, the ratio of first-order L-type lateral roots (1OL-type LRs) to total lateral roots, the length of 1OL-type LRs, and the branching density of second-order lateral roots in the middle and apical sections of the deepest nodal root were enhanced in NERICA 4 under adequate P levels. Additionally, P application enhanced the rate of vertical nodal root elongation in NERICA 4, promoting deep root growth and sustained dry matter production under drought conditions.

Overall, our results showed that N and P fertilization induced specific morphological changes in the root systems of the upland NERICA varieties, optimizing water uptake and improving drought adaptation. Depending on field conditions (e.g., the presence or absence of impediments to deep root growth), N and P applications can enhance upland rice productivity under drought stress.

3.3 Continued Variety Improvement and Variety Registration

We have been following the roadmap for the release of promising lines developed at the end of the SATREPS project and have taken steps to release two promising rice lines in Kenya as new varieties. These lines include IR124713-1, which carries the *Ctb1* gene that moderately enhances cold tolerance at the booting stage (Fig. 3.3), and IR124675-13, which is introgressed with the *Gn1a* and *WFP* genes associated with increased number of grains per panicle (Fig. 3.4). The process of releasing these promising lines as new

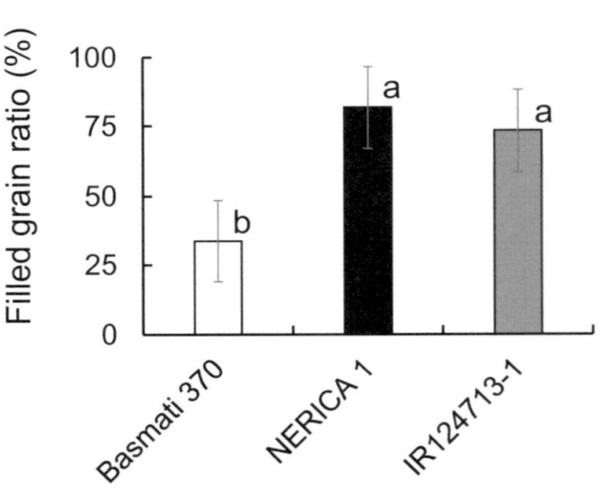

Fig. 3.3 Filled grain ratio (%) of Basmati 370, NERICA 1, and IR124713-1 (a Basmati 370 line introgressed with the cold tolerance gene *Ctb1*) during the low-temperature season in 2017 in Mwea, Kirinyaga County, Kenya. The filled grain ratio of IR124713-1 was significantly higher than that of Basmati 370, and comparable to NERICA 1, which is highly cold-tolerant. The same lowercase letters indicate no significant difference, as determined by the Tukey test

varieties began in 2021, delayed in part by the COVID-19 pandemic, which lasted for approximately 3 years, starting at the end of 2019.

Kenya has a well-established and strictly enforced system for variety registration compared to other sub-Saharan African countries, requiring significant time and budget. In 2021, as the impact of COVID-19 subsided, we submitted applications for the two lines to the Kenya Plant Health Inspectorate Service (KEPHIS), the government agency responsible for conducting the National Performance Trials (NPTs). The NPTs were conducted over two growing seasons from 2021 to 2023 at five sites across Kenya: paddy fields at KALRO Mwea, KALRO Mtwapa, National Irrigation Authority (NIA)-Bura, NIA-Ahero, and NIA-Bunyala (Fig. 3.5). Data on agronomic traits, pest and disease scores, and pre- and post-harvest traits were collected 28 days after transplanting (DAT), 56 DAT, 84 DAT, and at maturity. Furthermore, Distinctness, Uniformity, and Stability (DUS) tests were conducted at KALRO Mwea over two seasons from 2023 to 2024.

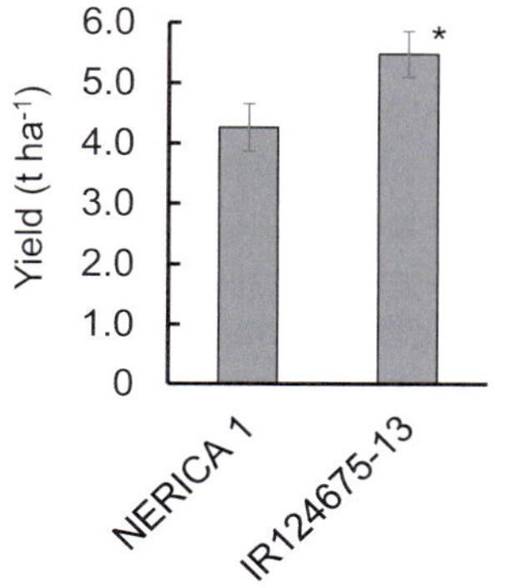

Fig. 3.4 Yield of NERICA 1 and IR124675-13 (a NERICA 1 line introgressed with the *Gn1a* and *WFP* genes associated with increased number of grains per panicle) grown in Mwea, Kirinyaga County, Kenya in 2020. The asterisk (*) indicates a significant difference from NERICA 1 based on the t-test

For IR124713-1, the two-season NPT data were reviewed in December 2023 by the National Performance Trial Committee (NPTC), which recommended the variety for release after comparing it to the check varieties. Subsequently, in December 2024, the National Variety Release Committee (NVRC) approved the line as a new variety named KALGOLD. For IR124675-13, additional NPT under upland conditions will be conducted for one season in 2025. The upland and irrigated lowland NPT data will be evaluated by the NPTC at the end of 2025, after which it is expected to advance to the approval process by the NVRC.

Fig. 3.5 Growth measurements conducted in July 2023 during the National Performance Trials in Bunyala, Busia County, Kenya

We have continued breeding efforts even after the conclusion of the SATREPS project and have developed a breeding line containing resistant alleles at *Pb1* and *Pi39* loci, which provide durable resistance to a wide range of blast races using MABC. The donor line C138 exhibited strong resistance in Mwea, and promising results were obtained from the blast resistance evaluation field at KALRO Mwea using BC_3F_3 generation. This blast-resistant Basmati 370 line, NU23_1971_30, is in the pipeline for NPT entry in 2025 following multi-location trials, together with PN1815, a Basmati 370 line introgressed with the *Gn1a* and *APO1* genes, which are associated with increased grain number per panicle. Currently, we are further advancing breeding efforts by incorporating genome editing to pyramid various beneficial genes, with the goal of achieving further improvements in rice varieties.

3.4 Seed Production Management Systems

KALRO is responsible for maintaining and managing the seeds of crop varieties developed by the institution. To achieve this, a department for seed multiplication and maintenance known as "KALRO Seeds" was established. As a result, KALRO is a registered seed merchant in Kenya, responsible for the production, processing, and marketing of seeds. Since rice is a scheduled crop for which seed certification is compulsory, as stipulated in the Seeds and Plant Varieties Act (Cap 326), the seeds produced at KALRO Mwea are inspected and certified by KEPHIS, the body responsible for seed certification and regulation in Kenya. At KALRO Mwea, both early generation seeds (EGS) and certified seeds are produced. EGS are produced in KALRO Mwea's agricultural fields, while certified seeds are produced by contracted farmers.

The seeds of the new rice varieties developed and released through KALRO Mwea are managed and produced by KALRO Mwea. However, because the seed system at KALRO Mwea is still under development, Nagoya University and KALRO Mwea collaborated to improve it through the aXis project conducted after the termination of the SATREPS project. In addition to installing the necessary facilities for seed production, the seed system for producing EGS was established at KALRO Mwea, and license agreements were signed with three seed merchants to enhance their capacity for producing certified seeds. Furthermore, the certified seeds produced were made available for sale through licensed seed merchants and agricultural input distributors, as well as directly to farmers through the KALRO Mwea Mkulima Shop.

As a result, the quantity of seeds produced has shown an impressive increase, rising from 1,738 kg in the 2019/2020 season to 57,000 kg in the 2022/2023 season (Table 3.1). While KALRO Mwea's seed production capacity has improved significantly, further enhancements in equipment, infrastructure, and human resources are necessary to meet the demands of producing seeds for the new rice varieties planned for release. Strengthening linkages with the rice seed value chain, including marketing and distribution, is also required.

3.5 Network Building, Human Resource Development

Even after the conclusion of the SATREPS project, the Japanese research team, led by Nagoya University, has continued collaborative research with KALRO Mwea by securing various external funding sources, thereby strengthening KALRO Mwea's functions as a research base. Through the SATREPS project and subsequent initiatives, KALRO Mwea's rice research environment was substantially enhanced, and research personnel were trained, leading to an increase in rice research activities. This progress has attracted considerable attention from universities within Kenya, international agricultural research organizations, and universities and agricultural research institutions in developed countries. In fact, new collaborative research projects involving KALRO Mwea and overseas research institutions have commenced, with some also involving Nagoya University. Furthermore, the rice research base at

Table 3.1 Quantity (kg) of early generation seeds and certified seeds produced at Kenya Agricultural and Livestock Research Organization Mwea during 2019/2020–2022/2023 seasons

Category	2019/2020	2020/2021	2021/2022	2022/2023
Early generation seeds	1,738	28,643	12,292	47,000
Certified seeds	0	10,687	19,537	10,000
Total	1,738	39,330	31,829	57,000

KALRO Mwea hosts local graduate students, providing research guidance for their degree programs. Thus, the rice research base at KALRO Mwea is establishing itself as a hub for the international rice research network in East Africa.

3.6 Prospects

The SATREPS project and subsequent collaborative research projects have generated a wide range of research outcomes, as outlined above. A promising line developed through the SATREPS project was approved as a new variety in Kenya and is nearing the stage where it can be distributed to farmers through the seed system at KALRO Mwea. While the Japanese members of the SATREPS project are not the best personnel to directly engage in agricultural extension activities to disseminate the new varieties and cultivation techniques in the partner countries, we are committed to facilitating the transfer of our research achievements—such as new varieties and cultivation techniques that maximize their potential—to farmers. This includes efforts to extend projects and seek new partnerships.

The SATREPS project was concluded successfully, achieving significant results in variety improvement and cultivation technology development. However, the development of new varieties and cultivation techniques is ongoing and must continue in the future. It must continuously adapt to scientific advancements, changes in cultivation environments, and evolving social, economic, market, and agricultural needs. For instance, it is necessary to improve varieties to cope with rapidly evolving blast races and to address new outbreaks of bacterial leaf blight. The increasing use of combine harvesters is likely to drive demand for rice varieties with lower shattering traits. As the rice market matures, the demand for high-quality rice is projected to increase significantly. Furthermore, the need for low-input, labor-saving rice cultivation techniques, as well as rice varieties suited to these practices, is greater than ever before. To respond swiftly to the rapidly changing environment surrounding rice cultivation, it is essential to incorporate cutting-edge scientific technologies such as genome editing.

The remarkable contributions through the SATREPS project and subsequent collaborative research projects have greatly improved the research environment at KALRO Mwea, fostered outstanding researchers, and strengthened its role as a base for rice research. KALRO Mwea's research base has already attracted international attention, and several international collaborative research projects are underway. Going forward, it is likely to attract numerous proposals and offers. However, due to a shortage of personnel, it is currently challenging to accept and implement all of these proposals. Therefore, it is crucial to continue enhancing KALRO Mwea's research capacity and actively developing human resources, building a stronger team of counterparts for international collaborative research. Once KALRO Mwea has significantly strengthened the quality and quantity of its research personnel, it could emerge as a hub for developing rice research talent, not only in Kenya but also across the broader East African region.

References

Angeles-Shim RB, Reyes VP, del Valle MM, Lapis RS, Shim J, Sunohara H, Jena KK, Ashikari M, Doi K (2020) Marker-assisted introgression of quantitative resistance gene pi21 confers broad spectrum resistance to rice blast. Rice Sci 27:113–123

Fukuta Y, Suzuki T, Yanagihara S, Obara M, Tomita A, Ohsawa R, Machungo CW, Hayashi N, Makihara D (2019a) Genetic variation of blast (*Pyricularia oryzae* Cavara) resistance in rice (*Oryza sativa* L.) accessions widely used in Kenya. Breed Sci 69:672–679

Fukuta Y, Telebanco-Yanoria MJ, Hayashi N, Yanagihara S, Machungo CW, Makihara D (2019b) Pathogenicities of rice blast (*Pyricularia oryzae* Cavara) isolates from Kenya. Plant Dis 103:3181–3188

Hasegawa T, Wainaina CM, Shibata A, Lucob-Agustin N, Makihara D, Kikuta M, Menge DM, Gichuhi EW, Samejima H, Kano-Nakata M, Kimani JM, Musila RN, Yamauchi A, Inukai Y (2022) The *outstanding rooting1* mutation gene maintains shoot growth and grain yield through promoting root development in rice under water deficit field environments. J Agron Crop Sci 208:815–829

Kikuta M, Menge DM, Gichuhi EW, Samejima H, Tomita R, Kimani JM, Musila RN, Doi K, Ashikari M, Angeles-Shim R, Jena KK, Makihara D (2023) Contribution of genes related to grain number (*Gn1a* and *WFP*) introgressed into NERICA 1 to grain yield under tropical highland conditions in central Kenya. Plant Prod Sci 26:309–319

Makihara D, Kimani J, Samejima H, Kikuta M, Menge D, Doi K, Inukai Y, Maekawa M, Masunaga T, Sasaki Y, Katsura K, Kitano H, Mitsuya S, Kano-Nakata M, Wainaina C, Gichuhi E, Njinju S, Kagito S, Magoti R, Kundu C, Yamauchi A (2018) Development of rice breeding and cultivation technology tailored for Kenya's environment. In: Crop production under stressful conditions: application of cutting-edge science and technology in developing countries. Springer, pp 27–47

Menge DM, Kameoka E, Kano-Nakata M, Yamauchi A, Asanuma S, Asai H, Kikuta M, Suralta RR, Koyama T, Tran TT, Siopongco JDLC, Mitsuya S, Inukai Y, Makihara D (2016) Drought-induced root plasticity of two upland NERICA varieties under conditions with contrasting soil depth characteristics. Plant Prod Sci 19:389–400

Menge DM, Onyango JC, Kano-Nakata M, Yamauchi A, Asanuma S, Tran TT, Inukai Y, Makihara D (2019) Effect of nitrogen application on the expression of drought-induced root plasticity of upland NERICA. Plant Prod Sci 2:180–191

Menge DM, Kano-Nakata M, Yamauchi A, Suralta RR, Makihara D (2020) Root and shoot responses of upland New Rice for Africa varieties to fluctuating soil moisture conditions as affected by different levels nitrogen fertilization. J Agron Crop Sci 206:322–337

Njinju SM, Hiroaki S, Keisuke K, Mayumi K, Joseph PG-O, John MK, Akira Y, Daigo M (2018) Grain yield responses of lowland rice varieties to increased amount of nitrogen fertilizer under tropical highland conditions in central Kenya Plant Prod Sci 21(2):59–70. https://doi.org/10.1080/1343943X.2018.1436000

Reyes VP, Angeles-Shim RB, Mendioro MS, Manuel MCC, Lapis RS, Shim J, Sunohara H, Nishiuchi S, Kikuta M, Makihara D, Jena KK, Ashikari M, Doi K (2021) Marker-assisted introgression and stacking of major QTLs controlling grain number (*Gn1a*) and number of primary branching (*WFP*) to NERICA cultivars. Plants 10:844

Samejima H, Katsura K, Kikuta M, Symon MN, John MK, Yamauchi A, Makihara D (2020) Analysis of rice yield response to various cropping seasons to develop optimal cropping calendars in Mwea Kenya Plant Prod Sci 23(3):297–305. https://doi.org/10.1080/1343943X.2020.1727752

Samejima H, Kikuta M, Katura K, Daniel MM, Emily WG, Cornelius MW, John MK, Inukai Y, Yamauchi A, Makihara D (2020) A method for evaluating cold tolerance in rice during reproductive growth stages under natural low-temperature conditions in tropical highlands in Kenya Plant Prod Sci 23(4):466–476. https://doi.org/10.1080/1343943X.2020.1777877

Samejima H, Katsura K, Kikuta M, Njinju Sm, Kimani JM, Yamauchi A, Makihara D (2021) Factors explaining differences in yield response to high nitrogen fertilization among rice varieties under tropical highland conditions in Central Kenya. JARQ 55(3):209–216. https://doi.org/10.6090/jarq.55.209

Tomita R, Reyes VP, Fukuta Y, Gichuhi EW, Kikuta M, Menge DM, Doi K, Makihara D (2023) Genetic variation of blast (*Pyricularia oryzae* Cavara) resistance in the Longistaminata chromosome segment introgression lines (LCSILs) and potential for breeding use in Kenya. Plants 12:863

Tsunematsu H, Yanoria MJT, Ebron LA, Hayashi N, Ando I, Kato H, Imbe T, Khush GS (2000) Development of monogenic lines of rice for blast resistance. Breed Sci 50:229–234

Wainaina CM, Makihara D, Nakamura M, Ikeda A, Suzuki T, Mizukami Y, Nonoyama T, Doi K, Kikuta M, Samejima H, Menge DM, Yamauchi A, Kitano H, Kimani JM, Inukai Y (2018) Identification and validation of QTLs for cold tolerance at the booting stage and other agronomic traits in a rice cross of a Japanese tolerant variety, Hananomai and a NERICA parent, WAB56-104. Plant Prod Sci 21:132–143

Yamada S, Kurokawa Y, Nagai K, Angeles-Shim RB, Yasui H, Furuya N, Yoshimura A, Doi K, Ashikari M, Sunohara H (2020) Evaluation of backcrossed pyramiding lines of the yield-related gene and the bacterial leaf blight resistant genes. J Int Coop Agric Dev 18:18–28

4 Conservation and Utilization of Wheat Genetic Resources in Afghanistan Expanded with the Homecoming Wheat Landraces Collected Half a Century Ago

Tomohiro Ban, Mujiburrahman Arifi, Aziz Ahmad Osmani, Ibrahim Hassani, and Ahmad Massoud Maqsodi

Abstract

Afghan wheat genetic resources collected half a century ago are being revitalized under the SATREPS project and incorporated into modern agricultural research and development efforts. These genetic resources, including landraces collected by Dr. Hitoshi Kihara in 1955, are helping to address the challenges posed by climate change, political instability, and food insecurity. The project has established a shuttle breeding system, shortened breeding cycles, and enhanced the development of high-yielding, resilient wheat varieties. Cooperative efforts have also improved the conservation of genetic resources of important crops, with more than 4,000 accessions currently conserved in Afghanistan's Genebank. Despite political changes, including the return of the Taliban to power in 2021, international cooperation and human resource development efforts have laid the foundation for sustainable wheat production. This chapter will explore the impact of these efforts on agricultural stability, recognize the importance of conserving genetic diversity, and highlight the need for continued support for research, infrastructure, and gender-inclusive approaches in the Afghan agrarian sector.

Keywords

Afghanistan food security · Climate change impact · Genebank conservation · International collaboration · Shuttle breeding system · Wheat landraces

T. Ban (✉)
Kihara Institute for Biological Research, Yokohama City University, Yokohama, Kanagawa, Japan
e-mail: tban@yokohama-cu.ac.jp

M. Arifi
Agricultural Research Institute of Afghanistan, Ministry of Agriculture, Irrigation and Livestock, Badambagh, Kabul, Afghanistan

A. A. Osmani
The Faculty of Agriculture, Kabul University, Kart-e-Char, Kabul, Afghanistan

I. Hassani
Directorate of Agriculture Irrigation and Livestock in Bamyan, Ministry of Agriculture, Irrigation and Livestock, Bamyan, Afghanistan

New Delhi, India

A. M. Maqsodi
Emergency and Resilience Programme, Food and Agriculture Organization of the United Nations, Kabul, Afghanistan

T. Nagamine et al. (eds.), *Plant Genetic Resources (PGR) for Sustainable Crop Production*, Sustainable Development Goals Series, https://doi.org/10.1007/978-981-96-7117-5_4

4.1 Introduction

With the world's population exceeding 8.1 billion by the end of 2024, the world is more keenly focused on the Sustainable Development Goals by 2030 to eliminate hunger, food insecurity, and all forms of malnutrition. Meanwhile, global boiling due to severe climate change is posing an ever-increasing threat to food production, with high temperatures, droughts, massive floods that submerge cultivated land, and unprecedented locust outbreaks from East Africa to South Asia. Furthermore, the global COVID-19 pandemic and frequent national and international conflicts are straining supplies of food, feed, fertilizers, and energy needed for production, making international scientific and technological cooperation for human security important for daily food and safe living in developing countries.

At the end of August 2021, the US administration completed the withdrawal of US troops stationed in Afghanistan, ending the 20-year-long Afghan war that erupted in the wake of the 11 September 2001 terrorist attacks in the United States. The Islamic fundamentalist Taliban, seeking a comeback, invaded the capital, Kabul, on 15 August 2001 while capturing major cities, leading to the collapse of the Islamic Republic of Afghanistan, which was established in 2004 after interim rule by the United States, the coalition of willing countries and the Northern Alliance. The withdrawal process was chaotic, and the news reports of the flooding of Kabul International Airport to evacuate not only the stationed troops and their citizens but also the families of officials and local staff involved in international cooperation shocked the world. Although the Taliban regime established a transitional government (the establishment of the Islamic Emirate of Afghanistan) on 27 September of the same year, it has not fulfilled its promises to protect human rights, particularly on women's issues, which disturbs cooperation with the international community. This section examines the wheat situation in Afghanistan, focusing on the background of the current situation in Afghanistan and the most recent online information.

4.2 "Crossroads of Civilizations": Afghanistan and Agricultural Development Cooperation

The difficulty for a nation to engage in development cooperation as the world faces the situation and food crisis in Afghanistan can be understood in terms of the historical inevitability of the geopolitical risks. In the nineteenth and twentieth centuries, Afghanistan became a British protectorate during the Great Game between the British and Russian empires for supremacy in Central Asia and the First and Second Afghan Wars in 1838 and 1878, and it is estimated that agricultural development cooperation for food production was also developed. During the Third Afghan War, the borders of India were finalized in 1919, and Afghanistan gained independence from the British in 1921. Afghanistan was born as a kingdom of Afghanistan on a modernization path with defined borders in 1926 for the first time in its history. During and after World War II, Afghanistan was exposed to the geopolitical risks of the struggle for supremacy between the major powers. It conducted balanced diplomacy between the East and West camps, and Japan, which established diplomatic relations with Afghanistan in 1934, sent Mr. Mitsuo Ozaki as a fruit and vegetable cultivation instructor from the Ministry of Agriculture and Commerce from 1935 to 1938 at the invitation of the Afghan government. He was later appointed to the Afghan government, paving the way for subsequent agricultural development cooperation (Ozaki 2003).

Thus, the country has prospered as a diverse complex of ethnic, cultural, and religious exchanges through repeated conflicts and disputes from its commissioning until today. It remains a multi-ethnic nation composed of Pashtuns (of Iranian origin), Tajiks (of Iranian origin), Uzbeks (of Turkic origin), Hazaras (of Mongolian origin), and others. The climate is Mediterranean and dry, and wheat, from which naan is made, is the staple food of most Afghans, accounting for more than 70% of their diet.

Domestic production and consumption of wheat, the staple food, have changed in response

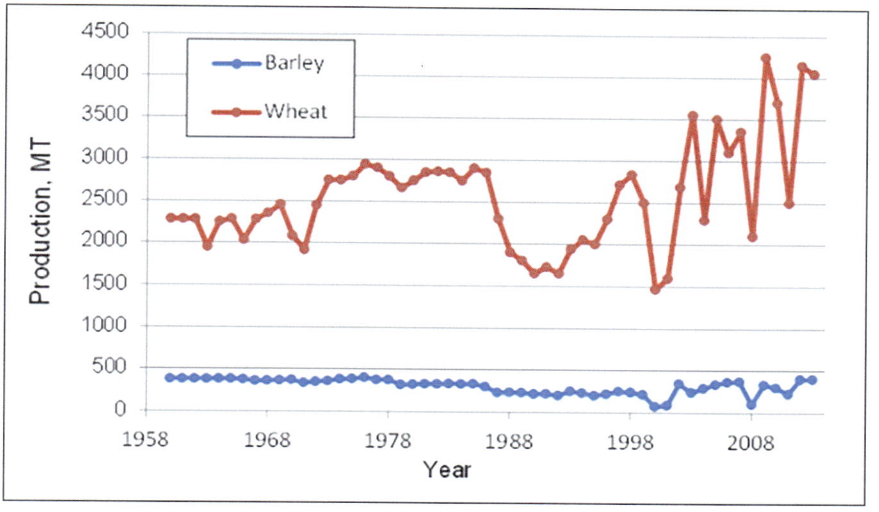

Fig. 4.1 Production of barley and wheat in Afghanistan from 1960 to 2013 (t). (Source: USDA Foreign Agriculture Service, Commodity Intelligence Report (2014))

to the regime's stability and repeated conflicts since 1960 (Fig. 4.1, Table 4.1). Domestic consumption of wheat until the mid-1980s was largely self-sufficient under the socialist regime, indicating that Afghanistan was an agricultural country (Figs. 4.1 and 4.2). In the spring of 2000, Eurasia experienced an unprecedented drought, and since then, the effects of climate change on wheat production have become apparent. In 2001, the Afghan War, in which the United States and the Northern Alliance attacked the Taliban in response to the Taliban's protection of al-Qaeda leader Bin Laden, who was responsible for the 9/11 attacks, turned Afghanistan into a war zone and destroyed wheat production (Fig. 4.1). However, domestic wheat consumption continued to grow under the regime of the Islamic Republic of Afghanistan (2004), which was established with the support of Western countries (Fig. 4.2). The intensification of civil war (2009–2011) caused by the largest deployment of troops from the International Security Assistance Force (ISAF) of the North Pacific Treaty Organization (NATO) disrupted civilian life and wheat production was temporarily destroyed (Table 4.1), but economic aid and international humanitarian assistance helped Afghanistan to become self-reliant. International cooperation increased wheat production (Table 4.1), but domestic consumption continued to grow, making it essential to double production. Japan also deployed comprehensive assistance that required the realization of international peace and human security for peacebuilding and ODA (Ministry of Foreign Affairs 2006). Thus, Afghanistan has been geopolitically at the mercy of the international community, and each time the social system has been reset by the escalating civil war, people's lives have collapsed, food security has been threatened, and the challenges of international agricultural development cooperation have been repeated.

4.3 Climate Change and Its Impact on Wheat Production In Afghanistan

4.3.1 Agro-ecological Zones and Wheat Production in Afghanistan: Climate and Market Dynamics

Afghanistan falls under the temperate winter rainy climate (i.e., Mediterranean-style climate) and steppe climate with dry summers and rainfall

Table 4.1 Area (ha), production (t), and unit yield (t/ha) of wheat in Afghanistan during 2011–2023

Market year	Area (1,000 ha)	Production (1,000 tons)	Yield (t/ha)
2011/2012	2100	2500	1.2
2012/2013	2512	5050	2.0
2013/2014	2550	5000	2.0
2014/2015	2560	5025	2.0
2015/2016	2550	5000	2.0
2016/2017	2300	5100	2.2
2017/2018	2104	4281	2.0
2018/2019	2000	3600	1.8
2019/2020	2300	5000	2.2
2020/2021	2330	5010	2.2
2021/2022	2200	4400	2.0
2022/2023	2200	4500	2.0
5-year average 2017/18–2021/22	**2187**	**4458**	**2.0**
Percent change from 5 year average (%)	**1**	**1**	**1**

Source: USDA Foreign Agriculture Service, IPAD (2022) PS&D Online updated on December 9, 2022

in winter, and in general, it has a harsh climate with hot summers and cold winters with a large temperature difference of about 30 °C (Kono 2019). The average annual precipitation for the entire land is 299 mm, and the grassland and southern areas fall under the desert in the climate classification by vegetation. There are important cities with population concentrations in each of the seven agro-ecological zones based on agro-meteorological analysis, comprising the main wheat production areas and consumer markets (Fig. 4.3). Kabul, the capital, is located in the center of the country and is the midpoint between the east, west, north, and south. Jalalabad supplies wheat to the eastern part of the country, while Mazar-e-Sharif supplies the northern provinces and, in good crop years, the southern provinces. Kunduz supplies the northeast and Faizabad, the chronically food-insecure Badakhshan province. The Maina market supplies the drought-prone Northwest. Herat supplies the west, and Kandahar supplies wheat to the southwest, where drought, civil unrest, and conflict often hamper markets (FEWS NET 2023).

4.3.2 Adapting to Droughts and Environment Challenges

Due to extremely low precipitation in Afghanistan, especially in summer, irrigated agriculture has been practiced using karez, which uses small streams of meltwater from lingering snow and underground water veins in the western, southern, and eastern provinces since the Bronze Age (Himat and Dogan 2019). Rainfed wheat is cultivated mainly in the northern states, with some areas producing higher yields than irrigated wheat production (Fig. 4.4). Recent climate change has increased drought in Afghanistan, and the drought damage in 2000 that hit all of Central Asia and Afghanistan caused a famine, and production declined sharply until the following year due to the lack of seed availability for sowing (Figs. 4.1 and 4.2). Thereafter, droughts occurred every two to four sowing seasons in 2004, 2008, and 2011, resulting in a sharp decline in both yield and production (Fig. 4.1, Table 4.1). Drought damage became more severe in the rainfed wheat regions, with a major drought in 2018 in the northwestern province displacing

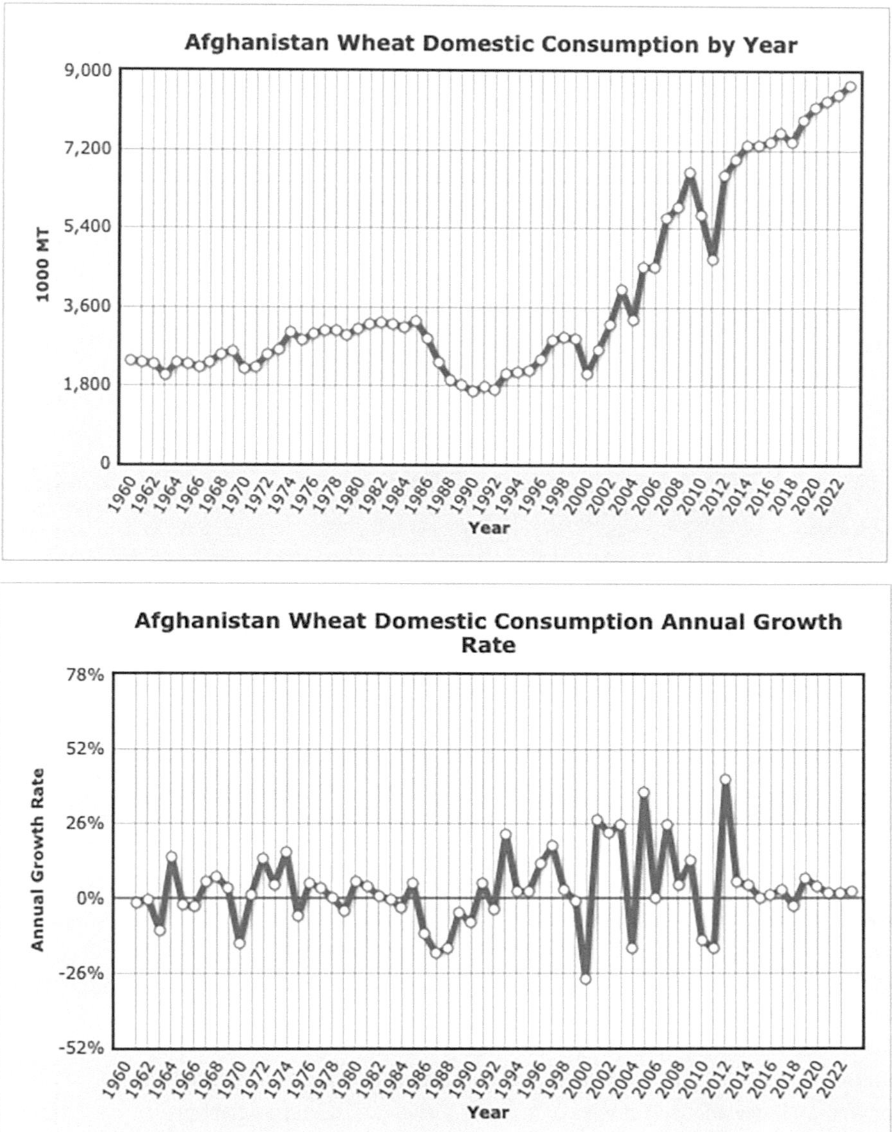

Fig. 4.2 Domestic consumption by year and annual growth rate of Afghanistan wheat from 1960 to 2022. (Source: IndexMundi (2024))

over 32,500 internally displaced persons (IDPs) from 7,000 families into urban areas (IOM 2019; Fig. 4.5).

There are four possible causes for the increase in drought in Afghanistan: (1) drought caused by the loss of residual snow in summer mountains due to long-term temperature increases and reduced spring snowfall, (2) reduced spring rainfall, (3) reduced winter snowfall on the plateau, and (4) increased evapotranspiration due to rising temperatures (Kono 2019). The average temperature in Afghanistan as a whole has increased by 1.8 °C from 1950 to 2010, about 2.5 times the global surface temperature increasing 0.71 °C, with increases since 2000 in the north, central highlands, and south. At the same time, precipitation in the central highlands has decreased by almost 40% by 2010, to the extent that evapo-

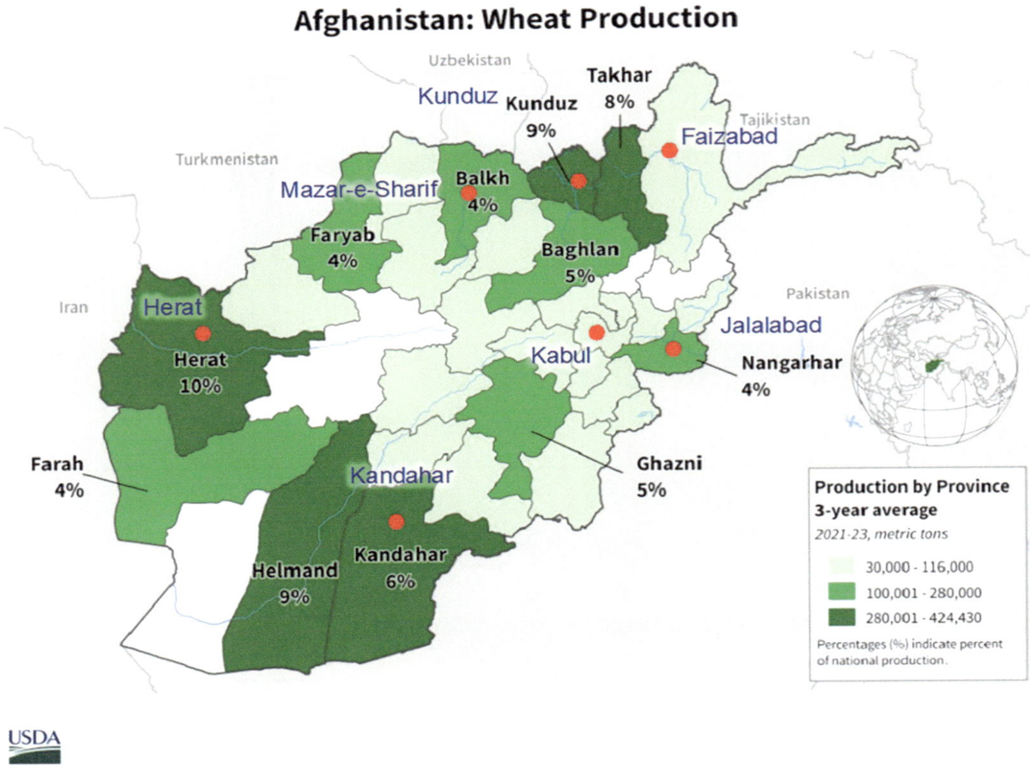

Fig. 4.3 Total wheat production and regional percentage production for the 2021–2023 season in Afghanistan. (Source: Modified from IPAD, USDA Foreign Agriculture Service (2023))

transpiration has declined over a wide area, from the southwest to the west. While drought conditions are a concern, Afghanistan's steep, mountainous terrain with few trees results in low water retention in mountainous areas, and strong spring rains cause flooding in large rivers. The decrease in precipitation in recent years has resulted in torrential rains that have scraped away fields and settlements in river basin farming villages, causing frequent flooding in urban areas. Annual fluctuations in wheat production have been extremely high since 2000 due to high temperatures and droughts linked to recent global climate change. The establishment of the Islamic Republic of Afghanistan, which was supported by Western countries, did not sufficiently improve agricultural and rural development infrastructure and socio-economic systems to achieve sustainable food production, which is important for doubling wheat production. Now that the second Taliban regime is in power, the multifaceted effects of climate change are compounded, and there are concerns about new difficulties and confusion in food production and agricultural development cooperation.

4.4 Agricultural R&D Cooperation and International Collaboration for Sustainable Wheat Production In Afghanistan

4.4.1 Revitalizing Wheat Genetic Resources Through International Collaboration

Changes in wheat productivity are influenced by political regimes, climate change, and the spread of new technologies. In Afghanistan, three trends can be observed: traditional agriculture before

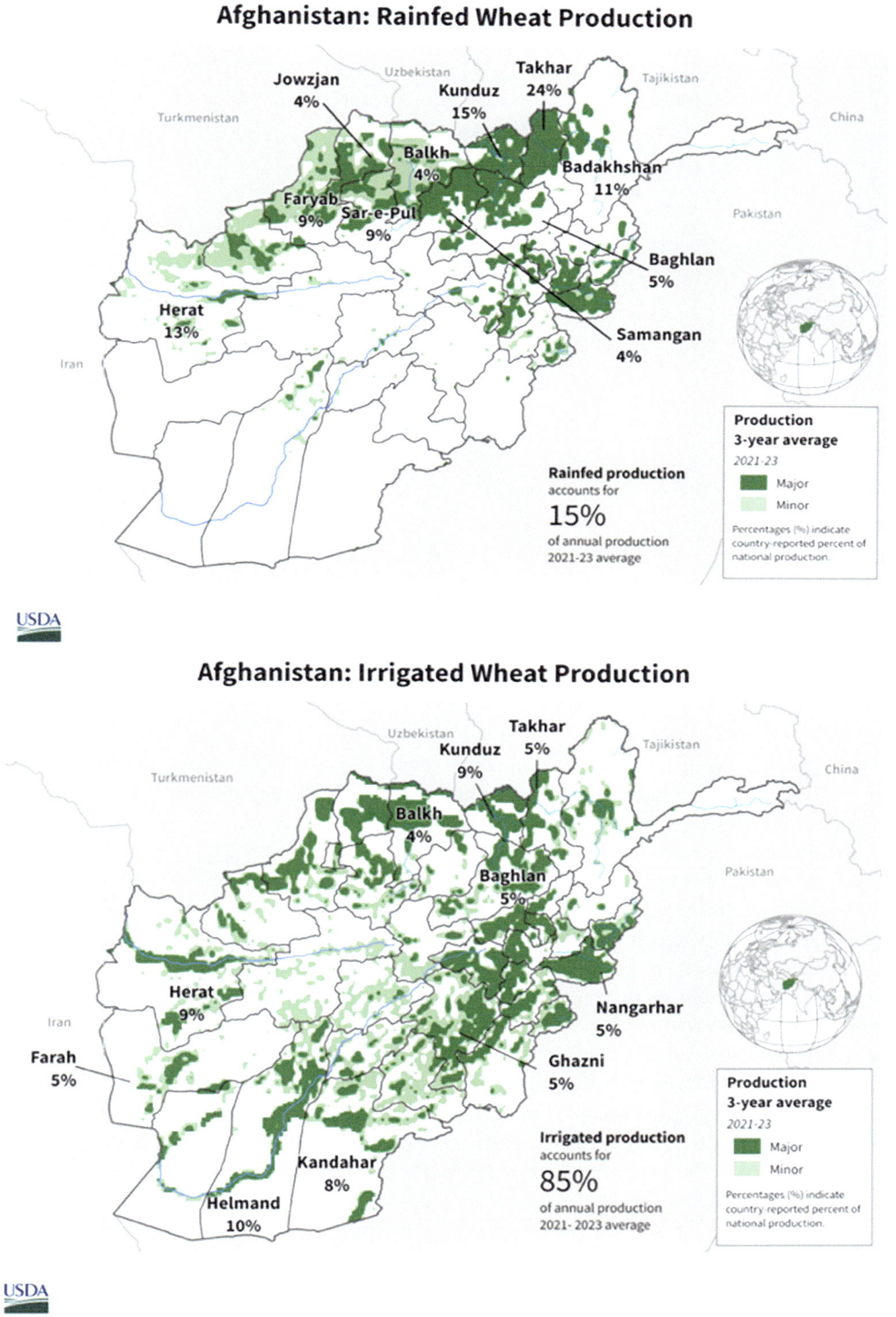

Fig. 4.4 Total wheat production and regional percentage production in Afghanistan for the 2021–2023 season under rainfed and irrigated cultivation. (Source: IPAD, USDA Foreign Agriculture Service (2023))

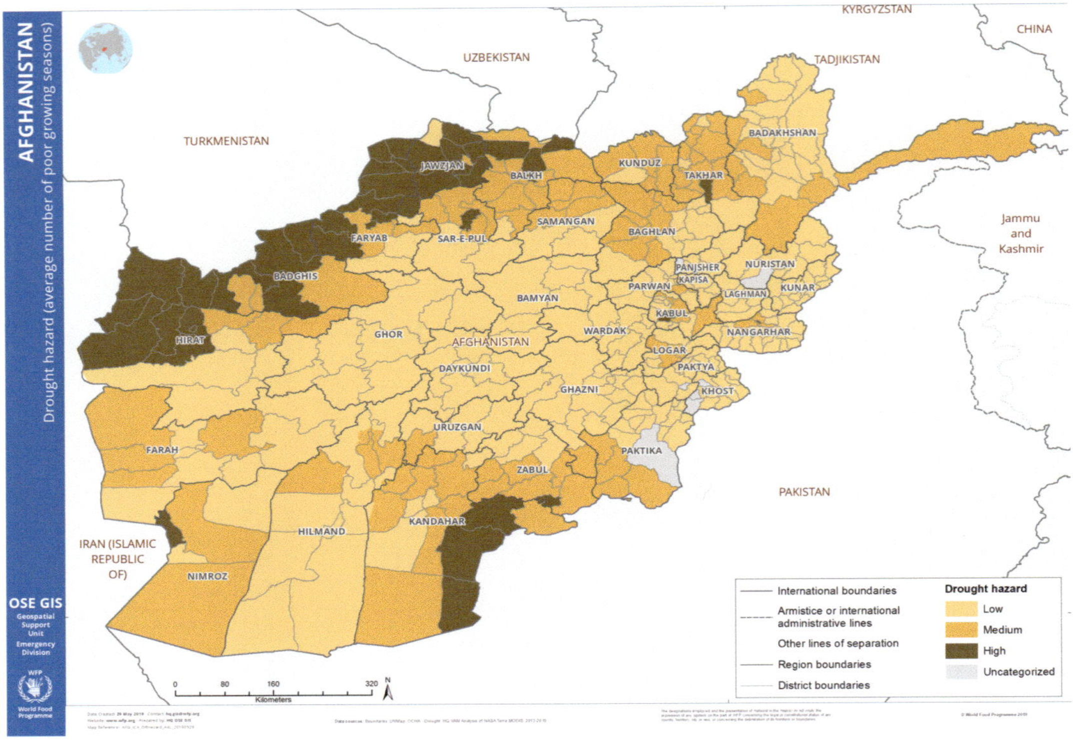

Fig. 4.5 The 2013–2018 drought hazard map of Afghanistan. (Source: FAO and IISA (2019))

the 1960s, stability due to the "Green Revolution" in the 1970s–1990s, and rapid increase in unit yield after 2000. Especially after 2000, international aid, the introduction of agricultural technology, and the development of irrigation facilities supported the expansion of production. The Marwarid irrigation canal completed by Dr. Tetsu Nakamura and others contributed to the agricultural development of the region. On the other hand, the diffusion of modern spring wheat varieties has increased the yield per unit of production, but the reduction in diversity and vulnerability to climate change have become issues.

From 2002 to 2016, spring wheat varieties developed by CIMMYT were tested for adaptability and disseminated in 11 locations in Afghanistan, resulting in an increase in spring wheat yield from 2.5 t/ha to a maximum of 5.98 t/ha. On the other hand, in 2015, these varieties accounted for more than 75% in four provinces. It reduced genetic diversity, and dependence on fertilizer and irrigation infrastructure became a challenge, leading to concerns about vulnerability to climate change and increased production variability.

4.4.2 The Role of SATREPS in Sustainable Agriculture

However, in areas where winter and facultative wheat are produced under rainfed cultivation, semi-dwarf varieties are susceptible to drought, resulting in unstable production. To address this, the SATREPS project "Development of Wheat Breeding Materials for Sustainable Food Production" from 2010 to 2015 conducted research on wheat genetic resources suitable for the natural environment of Afghanistan, improved varieties, and trained young researchers (Ban 2018). The project utilized native Afghan wheat landraces collected by Dr. Hitoshi Kihara and others in 1955 to provide locally adapted breeding materials. SATREPS took leadership for con-

Fig. 4.6 Genebank activities at ARIA MAIL in Afghanistan. (**a**) Herbarium of KAWLR collection, (**b**) building a new seed vault and laboratory in ARIA, (**c**) wheat landrace collection in ARIA

servation and its utilization of the traditional genetic resources for wheat breeding in Afghanistan, expanded with the 360 accessions of the homecoming wheat landraces (KAWLR collection) collected half a century ago (Fig. 4.6a). In collaboration with international organizations, the project also laid the foundation for sustainable agricultural development that takes advantage of the diversity of genetic resources.

4.4.3 Present Status of Genebank and Conservation of Plant Genetic Resources in Afghanistan

Conservation of plant genetic resources remains one of the most important units of the Afghanistan Agricultural Research Institute (ARIA), Ministry of Agriculture, Irrigation and Livestock (MAIL), as it is one of the most important components of the crop breeding and research and development program. SATREPS has been one of the key projects to make this unit functional and has provided all the necessary equipment and technical capacity building for its development. Following the example of the conservation and breeding use of wheat genetic resources, the unit is currently conserving and managing about 4,000 accessions of genetic resources from various important crops (Table 4.2). To maintain and evaluate these genetic resources and utilize them in the country, the budget and support provided by the current government and development projects are insufficient. Their practical and basic research activities for the development of new varieties are not very active. The return to power of the Taliban and their subsequent dominance have significantly disrupted the operations of most administrative units in Afghanistan. Many have either ceased functioning or slowed down considerably. One such unit is Afghanistan's Genebank, which in recent years has been instrumental in preserving plant genetic resources such as wheat, chickpeas, barley, maize, and medicinal plants.

Table 4.2 Accessions of plant genetic resources conserved in Genebank of Agricultural Research Institute, MAIL of Afghanistan

Crop entry#	Name of crops	Number of accessions	Crop entry#	Name of crops	Number of accessions
1	Wheat collection from Herat province	2,427	20	Lettuce	3
2	Wheat landrace collection from the north region and Kabul	57	21	Carrots	7
3	Improved bread wheat	219	22	Pepper	7
4	Durum wheat	39	23	Spinach	3
5	Triticale	39	24	Eggplant	8
6	Kihara Afghan Wheat Landrace (KAWLR) SATREPS, Japan	360	25	Coriander	9
7	Barley	199	26	Radish	10
8	Rice	45	27	Cucumber	6
9	Millet	3	28	Squish	6
10	Maize	44	29	Watermelon	15
11	Cotton	51	30	Okra	2
12	Sesame	51	31	Turnip	7
13	Soybean	10	32	Onion	15
14	Pea	292	33	Basil	8
15	Bean	10	34	Tomatoes	13
16	Mung bean	10	35	Sugar beet	6
17	Pigeon pea	2	36	Sorghum	5
18	Broad bean	3	37	Clover	22
19	Leek	3		Total	4016

However, this vital institution has not been immune to the challenges and disruptions brought about by the recent political changes.

Since 2011, when the SATREPS project became operational, the importance of rediscovering and revitalizing the value of the genetic resources of wheat landraces in the country through education and projects has been recognized. Afghanistan's Genebank has shown remarkable results, with 2,844 wheat landrace accessions (crop entry 1, 2, and 6), compared to other departments in MAIL (Fig. 4.6c, Table 4.2). Experts have successfully preserved, revitalized, and utilized plant genetic resources collected decades ago. These activities have been extended to the entire state with diverse climatic conditions, including Jalalabad, Bamiyan, Herat, and Mazar-e Sharif, preserving the valuable wheat landraces as the core of Genebank (Table 4.2).

However, the change in government resulted in a substantial disruption or complete halt on the research and conservation activities of Genebank. The factors contributing to this disruption include the following:

1. An exodus of skilled professionals and a shortage of human resources for the collection and preservation of genetic resources.
2. A lack of funds to sustain research activities even though the current administration is building a new seed vault (Fig. 4.6b).
3. Neglect by administrative authorities of the importance of research in collecting, conserving, and utilizing plant genetic resources.
4. Discontinuation of scientific research cooperation among universities, the Ministry of Agriculture, research institutes, and other organizations that have been active in agricultural research.

4.4.4 Establishment of a Shuttle Breeding System and Its Impact on Wheat Breeding in Afghanistan

In Afghanistan, around 80% of the people are engaged in agriculture, and the conservation and breeding use of plant genetic resources is an essential basic unit to support sustainable food production. Breeding of wheat, a staple food, is a national project of ARIA, the central agency of MAIL, and is also continued by each provincial agency, but activities are limited due to financial difficulties. ARIA is currently moving from traditional breeding systems to basic research for the introduction of advanced breeding methods, with direct involvement of affiliated local research institutions and NGOs. The government is streamlining efforts to promote private sector business development in the agricultural sector and public institutions to promote research and informational programs on environmental adaptability to rural agro-climates.

The national breeding program for the genetic improvement of wheat in Afghanistan is also very important for the development of new high-yielding and resistant varieties with high adaptability to a wide range of ecosystems. Unfortunately, this direction was missed, and Afghanistan had not been able to develop high-yielding commercial varieties with a wide range of adaptability that introduced multiple pest and environmental stress tolerance. Therefore, the SATREPS Afghan wheat project has launched a shuttle breeding platform to accelerate the breeding of rainfed wheat varieties based on the characteristics of wheat's unique genetic resources. Shuttle breeding, a breeding method developed at CIMMYT in Mexico, has accelerated the years of development of commercial wheat varieties and contributed significantly to sustainable food production in developing countries (Ortiz et al. 2007).

The SATREPS project organized a meeting to promote the shuttle breeding system, brought together local wheat breeders from all over Afghanistan, and launched the breeder's network and a framework of environmental adaptability trial using a unified methodology. Subsequently, wheat shuttle breeding was initiated between Kabul and Jalalabad, as well as between Kabul and Bamiyan. This platform considered research and development of specific breeding methods and human resource development to implement them, with the following five objectives:

1. Develop new types of wheat breeding material through a shuttle breeding system in Afghanistan to cut the time required for conventional breeding in half
2. Identify suitable genetic resources to serve as parents for future breeding programs
3. Develop complex multiple breeding programs using diverse genetic resources with high yield, disease resistance, and broad adaptability
4. Study the distinctive associations between wheat yield and yield-contributing traits
5. Incorporate wheat-specific landraces that are tolerant to biotic and abiotic stresses

Overlapping these goals, Afghan scientists from the SATREPS project are developing and maintaining segregating populations to meet breeding objectives and developing practical new high-yielding varieties through yield trials, multi-site evaluation trials, selection of promising lines, and the development of breeding materials.

For example, F_1 seeds are obtained by crossing of the modern varieties with Kihara Afghan Wheat Landrace (KAWL) from Japan at a warmer test site in Jalalabad, Nangarhar (Fig. 4.7a). F_2 seeds are sent to Bamiyan where a cooler area of the country in the same year to advance the generation and backcrossing (Fig. 4.7b). The F_3 lines of the segregating generation will be grown at the test site in Jalalabad and the ARIA field in Kabul to select rust-resistant lines and will be continuously repeated in Barmyan (Fig. 4.7c, d). Then, a group of productive and practical lines are selected for multi-site evaluation trials after the F_7 generation (Fig. 4.7e, f), and promising lines are selected from yield trials of the F_9–F_{10} generation, which took 6 years under the previous regime (Fig. 4.7g). Although the breeding program was stalled due to the restrictions of COVID-19, a new operation for developing breeding materials utilizing genetic resources resumed last year under the current administration. They are developing practical new high-yielding varieties with 64 lines of F_6, 15 lines of F_7, and 15 elite lines of F_8 generations in the 2024 cropping season.

Fig. 4.7 Overview of the wheat shuttle breeding system launched in Afghanistan. (**a**) F_1 cross in Nangarhar state, (**b**) advance F_2 generation and backcrossing in Bamiyan, (**c**, **d**) screening of Yellow and Brown rust disease resistance in Jalalabad and Bamiyan, (**e**, **f**) multi-site evaluation and yield trials, (**g**) propagation of promising lines for the release of a new wheat variety

4.5 Human Resource Development and Education for Women in Wheat Breeding and Genetic Resources

Capacity building of young Afghan researchers, which was the goal of the SATREPS project, is the most important element of the social implementation of the results of international collaborative research. Counterpart members of the ARIA MAIL research team collaborated to understand the importance of biodiversity and the conservation and sustainable use of wheat genetic resources. To gain experience in technology development for sustainable wheat production focusing on the interaction of genotype × environment × management in wheat breeding, a shuttle education program between Japan and Afghanistan was conducted through international cooperation with the Mexican and Turkish governments and international agricultural research institutions, such as CIMMYT, ICARDA, and FAO (Poole et al. 2022). The six young researchers from ARIA MAIL brought their skills, knowledge, and experience in genetic resources research back to Afghanistan after obtaining master's degrees in Japan. They contributed to the establishment of a facility for the Research Division of Plant Genetic Resources (RGRD), which serves as a gene bank for Afghan wheat and other crop genetic resources, stored and used for breeding material development (Table 4.2). However, in August 2021, the Islamic Republic of Afghanistan, which had been established with the support of Western countries, collapsed, and some of the students and local staff who grew up in the SATREPS project escaped to other countries, while many others survived the chaotic period by hiding with their families, fearing threats and life risks from the Taliban. During this second Taliban regime, the project was shut down for more than 6 months without salaries, but agricultural experimentation and research, the basis of their livelihood, has resumed, contributing to the reconstruction of Afghanistan's agriculture. Some of our counterparts who received their master's degrees in Japan returned to Afghanistan to teach at Kabul University and develop the SATREPS project research with their students. They are now studying in Germany for their PhDs, but their future path to return and develop their country is uncertain.

One of the most worrisome issues in Afghanistan today is the discrimination against women due to dogmatic interpretations of fundamentalism and human rights due to the suspension of girls' education. In developing countries, women face various constraints despite their large contribution to the rural economy (FAO 2011). Real revitalization and reconstruction of citizens' lives requires addressing gender disparities, which are also obstacles to agricultural development and food security. Data show that the revival of women's social inclusion and higher education, which was also addressed by the previous government in Afghanistan, and agricultural activities through the independence of female heads of households in rural areas can improve the structure of food crises (FSAC 2022). It is reported that women's work, primarily taking place in the home, as in rural Afghanistan, is invisible. However, women add value at various stages in the wheat value chain, including seed storage, cleaning and selection, cultivation, and harvesting operations. The research emphasizes the importance of recognizing their labor and contributions, as well as documenting their roles (Osmani et al. 2022). They argue for targeting and knowing agricultural strategies and programs as active participants and beneficiaries of women's agricultural research, extension, and priority setting. The Taliban regime has turned the former Ministry of Women's Affairs into the Ministry of Good and Evil, and the Ministry of Higher Education has suspended education for girls from secondary to university. Furthermore, in December 2022, women's work in international NGOs has been banned and suspended, and their activities have been halted. The Taliban regime is not recognized by the international community, which limits the activities of NGOs and international organizations working for peace and human security. If Afghanistan is not to be isolated from the international community, it must overcome this conundrum with regard to agricultural development cooperation.

4.6 Conclusion

Afghanistan has undergone repeated cycles of conflict and reconstruction since the beginning of time, and its flag has changed 18 times in the last 100 years (Al Jazeera 2021). The people living there have connected and developed their livelihoods in the same way while repeatedly embracing diversity. The wisdom to face the food crisis is concentrated in the genetic diversity of traditional wheat landraces that support the food and life base. Stability in Afghanistan will contribute not only to the reconstruction of the country but also to sustainable food production across the entire region, including food security in Central Asian countries with similar living conditions, as well as to the stability of the international community as a whole. Taking Tajikistan as an example, a neighboring country in the north that shares the same environment, ethnicity, and culture, it is dependent on Russia and Kazakhstan for wheat and wheat flour, its staple food, as well as for high-quality seeds, chemical fertilizers, and pesticides. They have suffered from the pandemic, the new coronavirus, and the global climate crisis. In addition, economic sanctions against Russia for its invasion of Ukraine have caused resource import prices to soar, greatly threatening the stability of the country. In response to global issues, we would like to explore solutions to the food crisis related to wheat shared by the nations of the Commonwealth of Independent States (CIS) in Central Asia and Eurasia as a whole. We also seek to consider the shift from national security to human security and research on environmentally regenerative agriculture to achieve sustainable production.

References

Al Jazeera (2021) Infographic: Afghanistan's flag over the years. https://www.aljazeera.com/news/2021/8/19/infographic-what-afghanistans-new-flag-looks-like-interactive. Accessed 14 Dec 2024

Ban T (2018) Development of wheat breeding materials using genetic resources in Afghanistan. In: Kokubun M, Asanuma S (eds) Crop production under stressful conditions. Springer, Singapore, pp 87–110. https://doi.org/10.1007/978-981-10-7308-3_6

FAO (2011) The state of food and agriculture, 2010–2011. Women in Agriculture: closing the gender gap for development. ISBN 9789251067680. https://openknowledge.fao.org/items/03d4e11d-8d41-4b98-9d3d-9cbb2f383d0a. Accessed 14 Dec 2024

FAO and IISA (2019) The Islamic Republic of Afghanistan – agro-ecological zoon atlas. Part 1: agro-climatic indicators. Roma. ISBN 978-92-5-131925-3. https://www.fao.org/geospatial/resources/detail/en/c/1252261/. Accessed 14 Dec 2024

FEWS NET (2023) Afghanistan Price Bulletin. https://reliefweb.int/report/afghanistan/afghanistan-price-bulletin-december-2022. Accessed 14 Dec 2024

FSAC (2022) Seasonal Food Security Assessment (SFSA) 2021 report. The Food Security and Agriculture Cluster (FSAC). https://fscluster.org/afghanistan/document/seasonal-food-security-assessment-sfsa-1. Accessed 12 Jan 2023

Himat A, Dogan S (2019) Ancient Karez system in Afghanistan: the perspective of construction and maintenance. Acad Platform J Sci 7(3):347–354. https://doi.org/10.21541/apjes.466757

IndexMundi (2024) Afghanistan wheat domestic consumption by year. https://www.indexmundi.com/agriculture/?country=af&commodity=wheat&graph=domestic-consumption.Accessed 14 Dec 2024

International Organization for Migration (2019) Displacement Tracking Matrix (DTM), Afghanistan – drought response situation report (01 March 2019). https://dtm.iom.int/reports/afghanistan-%E2%80%94-drought-response-situation-report-01-march-2019. Accessed 12 Jan 2023

Kono H (2019) Drought and floods in Afghanistan – the impact of climate change. TENKI 66(12):3–13 (in Japanese). https://doi.org/10.24761/tenki.66.12_773

Ministry of Foreign Affairs of Japan (2006) Evaluation of Japan's efforts toward peacebuilding: the case of Afghanistan – report (in Japanese). https://www.mofa.go.jp/mofaj/gaiko/oda/shiryo/hyouka/kunibetu/gai/afghanistan/jk05_01_index.html. Accessed 14 Dec 2024

Ortiz R, Trethowan R, Ferrara GO, Iwanaga M, Dodds JH, Crouch HH, Crossa J, Braun HJ (2007) High yield potential, shuttle breeding, genetic diversity, and a new international wheat improvement strategy. Euphytica 157:365–384. https://doi.org/10.1007/s10681-007-9375-9

Osmani AA, Ajir GM, Karimi PA, Madadi AK, Mayar N, Orwing M, Benson CN (2022) Mapping gender roles in family farm wheat production in central Afghanistan: evaluating roles, constraints, and opportunities. In: Gender perspectives on international development, vol 1. Michigan State University Press, pp 115–138. https://muse.jhu.edu/article/866197. ISSN 1947-4776

Ozaki M (2003) Afghanistan in the '30s as seen by the Japanese. Stone & Wind, Fukuoka (in Japanese), p 312. ISBN 4-88344-100-8

Poole N, Sharma R, Nemat OA, Trenchard R, Scanlon A, Davy C, Ataei N, Donovan J, Bentley AR (2022) Sowing the wheat seeds of Afghanistan's future. Plants People Planet 4:423–431. https://doi.org/10.1002/ppp3.10277

USDA Foreign Agriculture Service (2014) Afghanistan MY2014/15 winter grains northern rain-fed region concern due to low precipitation. https://ipad.fas.usda.gov/highlights/2014/01/Afghanistan/. Accessed 14 Dec 2024

USDA Foreign Agriculture Service, IPAD (2022) Afghanistan wheat area, yield and production. https://ipad.fas.usda.gov/countrysummary/default.aspx?id=AF&crop=Wheat. Accessed 14 Dec 2024

USDA Foreign Agriculture Service, IPAD (2023) Country summary, Afghanistan. https://ipad.fas.usda.gov/countrysummary/Default.aspx?id=AF. Accessed 14 Dec 2024

5 Wheat Breeding and Improved Technology Dissemination in the Hot and Dry Agroecosystems of Sudan

Hisashi Tsujimoto, Yasir S. A. Gorafi, and Izzat S. A. Tahir

Abstract

Global warming is occurring faster than expected, and its negative impact on crop production is becoming apparent. The impact on wheat production worldwide, particularly in Africa, is significant. Therefore, the rapid development and dissemination of climate-resilient varieties and technologies are urgently needed. In this chapter, we outline the development and identification of high-temperature-tolerant wheat genetic resources from wild species, the elucidation of the loci for tolerance through genetic analysis, the application of marker-assisted selection breeding, and the elucidation of the tolerance mechanism through physiological and biochemical analyses of tolerant wheat lines. In addition, a multi-stakeholder innovation platform approach and strategic public-private partnerships have been adopted to quickly and efficiently disseminate improved varieties and technologies to beneficiaries, which resulted in the production of substantial quantities of high-quality seeds and the enhancement of wheat farmers' productivity and profitability. These research and development activities were conducted in collaboration between Sudanese and Japanese researchers and other stakeholders.

Keywords

Heat and drought tolerance · Germplasm enhancement · Marker-assisted selection · Public-private partnership · Innovation platform

5.1 Introduction

Wheat is becoming an increasingly important food security crop in Africa, with consumption almost double the production across the continent. For instance, during the period from 2012 to 2022, Africa produced a total of 341.5 million metric tons (MMT) of wheat with an annual average of 26.3 MMT, whereas the total imported quantity was 674.8 MMT with an annual average of 51.9 MMT (FAOSTAT 2024a).

H. Tsujimoto (✉)
Arid Land Research Center, Tottori University, Tottori, Japan
e-mail: tsujim@tottori-u.ac.jp

Y. S. A. Gorafi
Graduate School of Agriculture, Kyoto University, Kyoto, Japan

Agricultural Research Corporation, Wad Medani, Sudan

I. S. A. Tahir
Arid Land Research Center, Tottori University, Tottori, Japan

Agricultural Research Corporation, Wad Medani, Sudan

T. Nagamine et al. (eds.), *Plant Genetic Resources (PGR) for Sustainable Crop Production*, Sustainable Development Goals Series, https://doi.org/10.1007/978-981-96-7117-5_5

Wheat has a long history in Sudan, with evidence of cultivation dating back thousands of years. The Nubians in northern Sudan were adept at cultivating wheat and used it as a primary food source. Wheat cultivation in Sudan has remained primarily confined to small areas north of latitude 17° N, using traditional cultural practices and varieties (Fig. 5.1). Due to steady population growth, ever-increasing urbanization, and shifts in dietary patterns, access to bread became an inevitable consequence of people's suffering because of its scarcity or unaffordable prices, which were beyond the reach of people with limited incomes.

Prior to World War II, all endeavors to extend wheat cultivation beyond the northern limits of Khartoum proved unsuccessful due to the absence of appropriate varieties and technologies. Since the 1960s, the Agricultural Research Corporation (ARC) has developed more than 70 different agricultural technologies, including more than 30 wheat varieties suitable for the heat-stress environment in Sudan. This resulted in wheat cultivation expansion into previously considered unsuitable areas, including Gezira, New Halfa, White Nile, and others (Fig. 5.1).

Despite the expansion of wheat cultivation and the genetic gain achieved through the development of heat-tolerant wheat varieties (Tahir et al. 2020a), the most significant challenge has been the considerable increase in consumption (Fig. 5.2). For instance, the annual rate of increase in wheat production from 1960 to 2020 was 9,571 tons, whereas the annual rate of increase in consumption during the same period was 45,961 tons (FAOSTAT 2024b). This equates to an annual increase in consumption that is more than four times the annual increase in production. To reconcile the discrepancy between wheat production and consumption, importing significant quantities of wheat has been necessary. For instance, 2.5 million tons were imported in 2014, with a

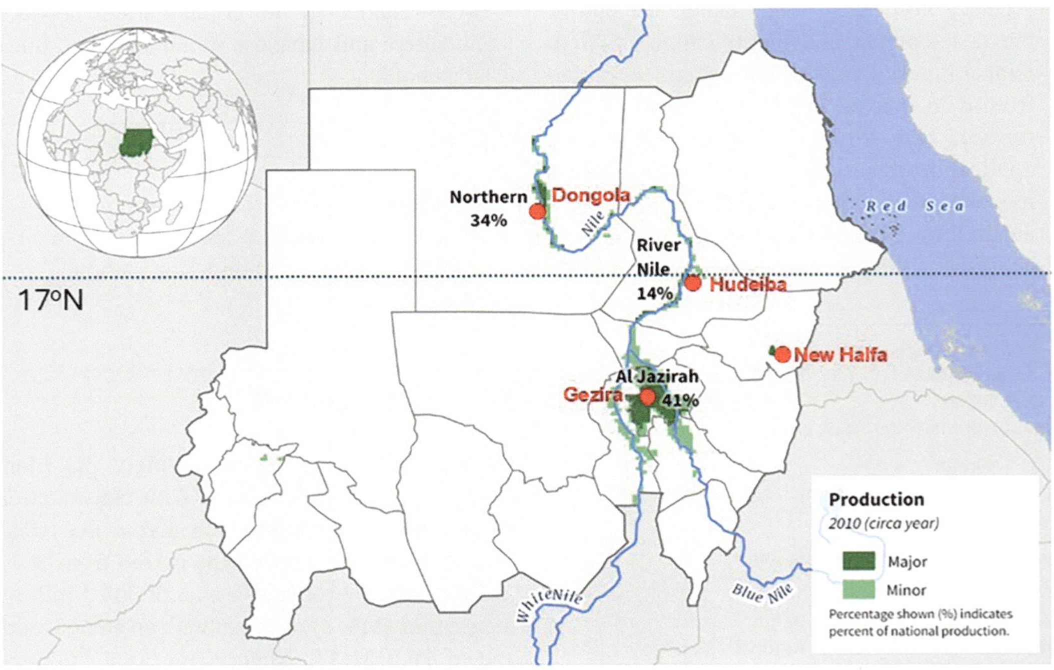

Fig. 5.1 Wheat production area in Sudan (green) and four agricultural research experimental stations in Sudan (red) where multi-locational trials were conducted. The dotted line is 17°N. The original map for the wheat growing area is from the USDA IPAD

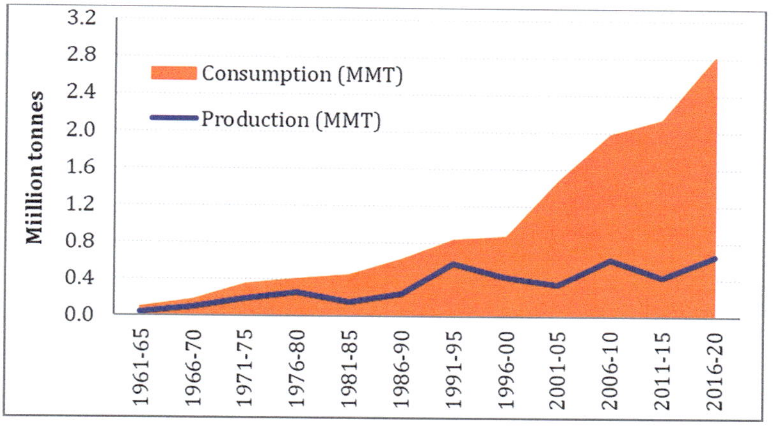

Fig. 5.2 Annual wheat consumption (orange area) and production (blue line) in Sudan

value exceeding one billion dollars. Therefore, it was necessary to adopt innovative approaches and techniques to enhance wheat productivity to address this challenge and bridge the significant gap between production and consumption in the face of climate change, which is likely to exacerbate this challenge further (Iizumi et al. 2021).

In this chapter, we describe the development of heat- and drought-tolerant wheat germplasm using new genetic resources, the application of marker-assisted selection based on genetic analysis of tolerance, the use of speed breeding technology, and the adoption of a multi-stakeholder innovation platform and strategic public-private partnerships to efficiently disseminate improved varieties and technologies to beneficiaries and to produce high-quality seeds to enhance the productivity and profitability of wheat farmers. This work was carried out as part of the activities of the SATREPS project "Sustaining wheat production through the development of climate-smart innovative technologies for the dry and heat-prone agroecologies of Sudan and sub-Saharan Africa," supported by the Japan International Cooperation Agency (JICA) and the Japan Science and Technology Agency (JST).

5.2 Germplasm Enhancement, Marker-Assisted Selection, and Speed Breeding

Climate change is outpacing the rate of genetic gain in breeding (Iizumi et al. 2022). Accelerated breeding is needed to develop varieties that can withstand climate change. Accelerated breeding involves speeding up a plant's growth cycle, which means the plant has to be grown smaller than it would be in the field. For this reason, traits such as heat and drought tolerance, which should be evaluated based on grain yield and different plant response characteristics, are challenging to assess in accelerated breeding facilities. In addition, tolerance is a quantitative trait with complex genetic behavior controlled by multiple genes. Therefore, when QTLs with a high phenotypic contribution are identified and tracked using linked DNA markers, marker-assisted selection (MAS) must also be applicable to quantitative traits. In wheat, intraspecific diversity is lower than that of other crops, and most of the promising QTLs have already been extensively used in modern breeding. Therefore, we have started to search for novel QTLs with high phenotypic effects in the gene pool using a genetically diverse wheat population.

Bread wheat (*Triticum aestivum*) is a hexaploid, originating from a natural hybrid between cultivated tetraploid wheat, which has A and B sub-genomes, and the wild species *Aegilops tauschii*, which has the D sub-genome. Since the number of individuals involved in this interspecific hybridization in the evolution was minimal, the wheat gene pool is primarily narrower than the diversity of other diploid crops that inherited the polymorphism from the ancestral wild species population. Later, gene introgression from cultivated tetraploid wheat somewhat increased the diversity of the A and B sub-genomes, but introgression of the D sub-genomes from *Ae. tauschii* did not occur due to cross-incompatibility between hexaploid wheat and *Ae. tauschii*. To enhance the diversity of the D sub-genome, we proposed the use of the "MSD (multiple synthetic derivative) population" (Tsujimoto et al. 2015). The MSD population is multigenic and has an intraspecific diversity of *Ae. tauschii* in a bread wheat cultivar "Norin 61 (N61)"(Elbashir et al. 2017). Later, Gorafi et al. (2018) randomly selected 400 MSD lines from the population, genotyped them by genome-wide DNA markers, and established an "MSD panel" for genetic analyses. Following the same concept applied in the bread wheat MSD populations, Balla et al. (2022a, b) selected diverse populations and lines of durum wheat (*Triticum durum*), which included wild tetraploid wheat genome segments.

Evaluating these bread and durum wheat populations in multi-environment trials revealed high genetic diversity for various stress-resilience traits. The genome-wide association study (GWAS) approach elucidated prominent heat and/or drought tolerance loci in the A, B, and D sub-genomes. We have produced near-isogenic lines (NILs) of bread wheat with one of the tolerant loci found in the A, B, or D sub-genomes to reveal detailed molecular and physiological properties of heat and drought tolerance (unpublished).

5.3 Evaluation of Diversity Panel in Stressed Environments

We evaluated all or a part of the 400 MSD lines in 16 different agroecological and management environments in Dongola, Hudaiba, and Gezira Research Station Farms, Sudan (Fig. 5.1). In Gezira (Wad Medani), during most wheat growing period, the maximum temperatures were higher than 30 °C and sometimes rose above 35 °C, especially towards the late grain filling period. Dongola was always 5–8 °C cooler than Gezira. At the Gezira Research Farm, we evaluated the MSD population under late sowing conditions to identify lines tolerant to excessive heat stress, under deficit irrigation to evaluate drought tolerance, and under low nitrogen and phosphorus conditions to identify lines with efficient nutrient use. Genotypes with enhanced tolerance to heat and water-deficit stress and more efficient resource use were identified (Gorafi et al. 2025). Grains harvested from heat, drought, and nutrient-deficient experiments were tested for nutritional and end-use quality attributes. Promising germplasm with novel QTLs, marker-trait associations (MTAs) and alleles, and improved nutritional and end-use quality under heat stress conditions were identified (Elhadi et al. 2021; Mohamed et al. 2022).

For example, Itam et al. (2022) selected 145 MSD lines with homogeneous heading times from the 400 MSD and tested them for drought tolerance at the Gezira Research Station Farm (Fig. 5.3a). Soil water potential was monitored every 2 h by sensors buried 20 cm in the soil. Drought was imposed by withholding irrigation when 50% of all genotypes in the drought treatment plots reached heading, while regular irrigation continued in the control plots. To avoid permanent wilting, the drought plots were rewatered when the soil water potential approached −900 kPa (Fig. 5.3b). It is worth mentioning that there is no rain during the winter season at the Gezira Research Station Farm, and relative humidity is generally low (20 to 30%). Morphophysiological traits were measured, and marker-trait association (MTA) analysis was conducted (Fig. 5.3c). The analysis revealed drought response MTAs on chromosomes 2D, 3D, 5D, and 7D.

Under field conditions, some of the MSD lines exhibited high levels of tolerance to environmental stresses compared to the recurrent parent, N61, and some Sudanese cultivars. This indicates their suitability as promising genetic resources

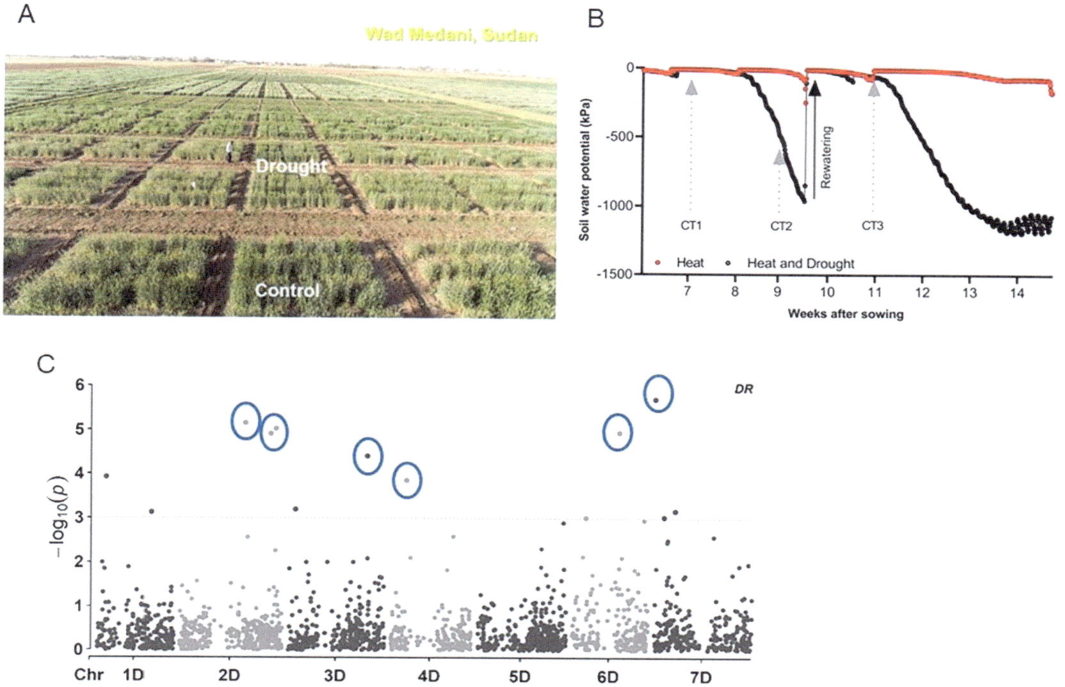

Fig. 5.3 Evaluation of MSD panel for stress tolerance. (**a**) Field experiment in Gezira Research Station. (**b**) Soil water potential in drought condition. (**c**) Manhattan plots for grain yield showing marker-trait association of drought tolerance

for enhancing stress tolerance and for physiological studies.

Several studies were carried out to elucidate the molecular and physiological mechanisms of heat and drought tolerance. For instance, Itam et al. (2020) conducted experiments in growth chambers in Japan, where the environment was adjusted to simulate the Sudanese field conditions. Three representative lines from the MSD panel were evaluated for drought tolerance with their recurrent parent N61, as a check, for 21 days at 22/18 °C day/night temperature. In the control treatment, soil moisture was maintained at the optimum field capacity. In the drought treatment, irrigation was withheld. When the soil moisture reached −1,000 kPa, photosynthetic parameters were measured using fully expanded leaves, and leaf metabolites and drought-responsive enzymes were analyzed in the harvested plant samples. The result showed that the three MSD lines responded to drought by inhibiting transpiration, closing stomata, and improving water use efficiency more than N61. In addition, metabolome analysis showed that all lines, including N61, accumulated general drought-responsive substances known as compatible solutes, such as proline and betaine, under drought stress conditions. However, one MSD line (MNH5) accumulated GABA and tartaric acid, while the other (MSD345) accumulated putrescine, adenine, and histidine, suggesting that these lines have novel, unknown mechanisms of drought tolerance in addition to the commonly known mechanisms (Fig. 5.4).

Chemical changes were monitored using Fourier transform infrared (FTIR) spectroscopy coupled with chemometrics in wheat plants subjected to heat stress at a daily maximum temperature of 42 °C for 3 days at the three-leaf stage (Osman et al. 2022a). The FTIR-based fingerprinting technique to characterize wheat leaves' metabolomic response to heat stress showed that

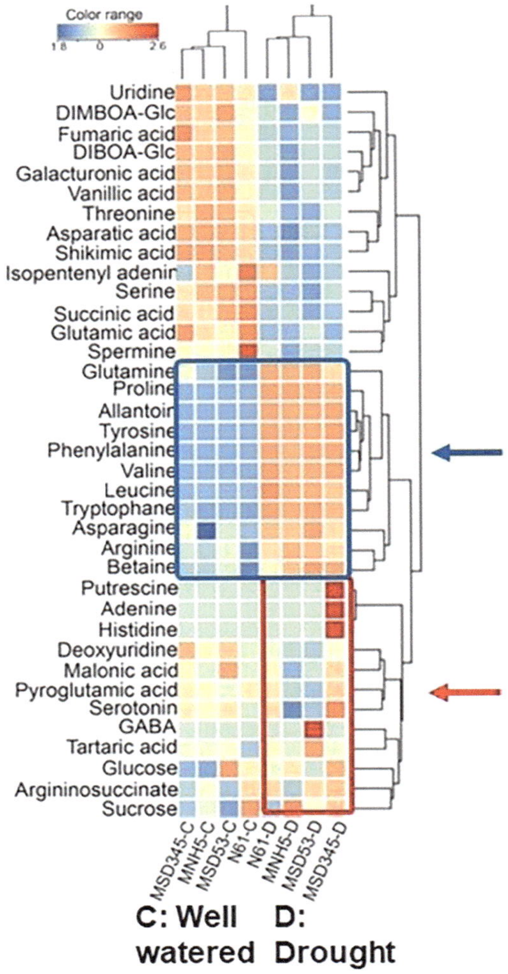

Fig. 5.4 Clustered heatmap of metabolite in leaves of the MSD lines and N61. Blue and red arrows indicate general and genotype-specific metabolites accumulated by drought stress, respectively

the linear discriminant analysis (LDA) discriminated between samples from stress and control conditions. Thus, LDA-based chemometrics using FTIR spectra provides a powerful approach for monitoring heat-induced chemical changes in wheat leaves. Using these discriminative wave numbers, novel spectrum-based biomarkers were developed that successfully diagnosed heat-stressed leaves (Osman et al. 2022a, b).

5.4 Microclimate Observations and Crop Modeling

Sustaining wheat production in Sudan is challenging due to the diverse climatic zones and increasing vulnerability to climate change. Sudan's wheat-growing areas, already among the hottest in the world, are expected to see their share of domestic production decline (Iizumi et al. 2021). Therefore, immediate adaptation strategies, such as breeding heat-tolerant varieties and optimizing sowing dates, are needed to mitigate some of the adverse effects and meet future demand, which requires annual yield gains of 0.2–2.7% (Iizumi et al. 2021).

Understanding the relationship between temperature and yield could enhance wheat productivity and resilience. The relationship between temperature over 48 growing seasons and irrigated wheat yields in three main wheat-producing regions was examined. Upward trends in annual and growing season maximum and minimum temperatures were found across all regions, with the Northern State showing the most pronounced increases (Musa et al. 2021). Several temperature indicators, especially minimum night temperatures and maximum temperatures during key growth stages (e.g., flowering, grain filling), were significantly and negatively correlated with grain yield. However, the strength of these relationships varied by region; minimum night temperatures and maximum day temperatures had the strongest negative correlations in the Northern and Gezira states, respectively, while both day and night temperatures were negatively correlated in Kassala State (Musa et al. 2021).

As the wheat crop grows as a large population in the field, the response of the population, as well as that of the individual plants and even specific parts of the plant, to the surrounding environment should be considered when breeding wheat for high-temperature tolerance. Therefore, in this project, microclimatic observations of irrigated wheat communities were carried out by installing an automated weather station inside the wheat field. For example, the results showed that the temperature at canopy level (0.6–0.8 m) was always cooler than that of 2 m above the ground, with a maximum difference of 8 °C, indicating that the population has a significant canopy cooling system. A pronounced evaporative cooling effect due to irrigation was also observed, and notably, canopy temperature reduction was different for different wheat cultivars with different phenology (Mohammed et al. 2023). Model simulation results indicated that the effects of high-temperature stress on yield could be more effectively modeled by using canopy temperature rather than air temperature (Iizumi et al. 2022). These microclimate and crop response findings were considered useful for crop modeling of soil-plant-atmosphere continuum systems in irrigated wheat fields under hot and dry conditions. This should be considered and incorporated into the wheat breeding and management programs in hot and dry agroecologies.

5.5 Establishing Innovation Platforms and Building Strategic Partnerships

Even if researchers develop new technologies, including new cultivars, after investing much effort and time, it does not necessarily mean they will be adopted and popularized. If successful strategies to address constraints to technology adoption are identified and solutions applied, increased technology adoption can contribute to economic growth and poverty reduction among farmers (Ibrahim et al. 2024). Prior to the implementation of project activities in Sudan, a study was conducted to assess the level of the adoption of improved technologies and management practices and to identify the key factors that influence their adoption in the major wheat-producing areas of Sudan. Access to quality seeds, financial credit, and extension services were the most critical determinants of farmers' adoption of improved technologies (Ibrahim et al. 2024).

Historically, the traditional linear model for research and extension was used to disseminate new technologies, including new varieties, to farmers. This model lacks the needed interaction and feedback learning among stakeholders along the crop value chain. To overcome this shortcoming and build strong relationships and interactions among all stakeholders, the integrated agricultural research for development and innovation system approach was introduced. Following this approach, the project established six innovation platforms (IPs) in four major wheat-producing areas in Sudan. The IPs allowed different stakeholders in the public and private sectors, including policymakers, to interact to understand each other better, learn and develop shared priorities, identify challenges, create opportunities and solutions, define roles, and agree on joint actions and work plans to realize agricultural revitalization and development (Tahir et al. 2020b). A series of activities were carried out within the IPs, including technology dissemination and scaling up, technical backstopping, capacity strengthening, seed production and distribution, field and harvest days, etc.

Securing high-quality seed is critical to sustainably producing and delivering essential food crops in required quantity and quality. Seeds are the foundation of any good agricultural practice that seeks to provide food security for an ever-growing population in the face of climate change. A sustainable and balanced strategy is needed to ensure continued, economically viable seed production and supply. Based on an earlier comprehensive analysis of the seed system, the project facilitated the establishment of strong public-private partnerships (PPPs) between public and private seed producers. Plans were made to produce breeder, pre-basic, and basic seeds of all commercial, heat-tolerant bread and durum wheat varieties and provide them to the public and private seed producers for certified seed production. These strong PPPs enabled the production of certified seeds of heat-tolerant wheat varieties and their dissemination to farmers at the IP sites and beyond. The quantity of seed produced through the project in partnership with the public and private sectors increased from 41.7 tons in the first season (2020) to 5,411 tons in the third season (2022) and to more than 13,600 tons in the fourth season (2023). These amounts represented only 0.12% of the total national seed production in the first season; however, the percentage increased to 1.5%, 18.7%, and 34% in the second, third, and fourth seasons, respectively. Even under extremely insecure conditions in 2024, the project, with its partners, produced more than 10,000 tons of seed for both commercial and recently released varieties.

A direct comparison between the participating farmers (PF) and nonparticipating farmers (NPF) in the project activities in 2023/2024 showed a 44.6% increase in the productivity of PF compared to NPF. For instance, farmers' feedback in the annual review and planting meetings and the field days indicated a considerable increase in areas and productivity of wheat at IP sites. Coupled with the capacity strengthening of the farmers and all other stakeholders along the wheat value chain, the increased productivity has increased the motivation of IP stakeholders and those outside the IPs and further increased the demand for high-quality seeds.

5.6 Future Prospects

The project started in 2018 to improve the nutritional value and accelerate the process of improving wheat to adapt to climate change by applying modern technologies and identifying the mechanisms that lead to wheat adaptation to abiotic stresses. The main activities included (1) evaluation of genetically diverse germplasm in multi-location, multi-environment field trials in Sudan for heat and drought tolerance for grain yield and other climate-resilient traits, (2) selection of stress-tolerant lines, (3) elucidation of the genetic behavior of heat and drought tolerance, and (4) introduction of marker-assisted selection for accelerated crop improvement. Future prospects include the use of promising lines identified in this project within Sudanese breeding programs; the establishment of research facilities to promote innovative and modern breeding technologies, including marker-assisted selection and accelerated speed breeding; and the establishment and operation of innovation platforms to promote the dissemination of innovative production technologies, including the adoption of new varieties. Human capacity building for young scientists was a central objective of the project.

Implementing the activities faced significant challenges during the project period due to the deteriorating security situation, the global COVID-19 pandemic, and the outbreak of civil war in the second half of the project period. It is the most important responsibility of all humanity to ensure that people born on this planet, no matter where they live, have access to the food they need. We must not stop trying to address the food production challenge, as this project has done.

References

Balla MY, Gorafi YSA, Kamal NM, Abdalla MGA, Tahir ISA, Tsujimoto H (2022a) Harnessing the diversity of wild emmer wheat for genetic improvement of durum wheat. Theor Appl Genet 135:1671–1684. https://doi.org/10.1007/s00122-022-04062-7

Balla MY, Gorafi YSA, Kamal NM, Abdalla MGA, Tahir ISA, Tsujimoto H (2022b) Exploiting wild emmer wheat diversity to improve wheat A and B genomes in breeding for heat stress adaptation. Front Plant Sci 13. https://doi.org/10.3389/fpls.2022.895742

Elbashir AAE, Gorafi YSA, Tahir ISA, Kim J-S, Tsujimoto H (2017) Wheat multiple synthetic derivatives: a new source for heat stress tolerance adaptive traits. Breed Sci 67. https://doi.org/10.1270/jsbbs.16204

Elhadi GMI, Kamal NM, Gorafi YSA, Yamasaki Y, Takata K, Tahir ISA, Itam MO, Tanaka H, Tsujimoto H (2021) Exploitation of tolerance of wheat kernel weight and shape-related traits from Aegilops Tauschii under heat and combined heat-drought stresses. Int J Mol Sci 22:1–21. https://doi.org/10.3390/ijms22041830

FAOSTAT statistics database (2024a) Food Balance Sheets. Available online: www.fao.org/faostat/en/#data/FBS. Accessed on 25 Dec 2024

FAOSTAT (2024b) Available online: https://www.fao.org/faostat/en/#home. Accessed on 25 Dec 2024

Gorafi YSA, Kim JS, Elbashir AAE, Tsujimoto H (2018) A population of wheat multiple synthetic derivatives: an effective platform to explore, harness and utilize genetic diversity of Aegilops Tauschii for wheat improvement. Theor Appl Genet 131:1615–1626. https://doi.org/10.1007/s00122-018-3102-x

Gorafi YSA, Tahir ISA, Tsujimoto H (2024) A multiple synthetic derivatives population for mining Aegilops tauschii traits and genes in a background of common wheat. J Plant Regist 19:e20424. https://doi.org/10.1002/plr2.20424

Ibrahim AMM, Hassan AO, Idris AAM, Gorafi YSA, Tsujimoto H, Tahir ISA (2024) Key factors influencing the adoption of improved wheat production technologies in the irrigated, heat-prone, arid environments of Sudan. Sustainability 16:6600. https://doi.org/10.3390/su16156600

Iizumi T, Ali-Babiker IEA, Tsubo M, Tahir ISA, Kurosaki Y, Kim W, Gorafi YSA, Idris AAM, Tsujimoto H (2021) Rising temperatures and increasing demand challenge wheat supply in Sudan. Nat Food 2:19–27. https://doi.org/10.1038/s43016-020-00214-4

Iizumi T, Tsubo M, Maruyama A, Tahir I, Kurosaki Y, Tsujimoto H (2022) High-temperature indicators for capturing the impacts of heat stress on yield: lessons learned from irrigated wheat in the hot and dry environment of Sudan. Clim Res 89:85–98. https://doi.org/10.3354/cr01709

Itam M, Abdelrahman M, Yamasaki Y, Mega R, Gorafi Y, Akashi K, Tsujimoto H (2020) Aegilops Tauschii introgressions improve physio-biochemical traits and metabolite plasticity in bread wheat under drought stress. Agronomy 10:1588. https://doi.org/10.3390/agronomy10101588

Itam MO, Mega R, Gorafi YSA, Yamasaki Y, Tahir ISA, Akashi K, Tsujimoto H (2022) Genomic analysis for heat and combined heat–drought resilience in bread wheat under field conditions. Theor Appl Genet 135:337–350. https://doi.org/10.1007/s00122-021-03969-x

Mohamed IES, Kamal NM, Mustafa HM, Abdalla MGA, Elhashimi AMA, Gorafi YSA, Tahir ISA, Tsujimoto H, Tanaka H (2022) Identification of Glu-D1 alleles and novel marker–trait associations for flour quality and grain yield traits under heat-stress environments in wheat lines derived from diverse accessions of Aegilops Tauschii. Int J Mol Sci 23. https://doi.org/10.3390/ijms231912034

Mohammed AAA, Tsubo M, Ma S, Kurosaki Y, Ibaraki Y, Tahir ISA, Gorafi YSA, Idris AAM, Tsujimoto H (2023) Micrometeorological comparison of canopy temperature between two wheat cultivars grown under irrigation in a hot environment in Sudan. Agronomy 13:3032. https://doi.org/10.3390/agronomy13123032

Musa AII, Tsubo M, Ali-Babiker IEA, Iizumi T, Kurosaki Y, Ibaraki Y, El-Hag FMA, Tahir ISA, Tsujimoto H (2021) Relationship of irrigated wheat yield with temperature in hot environments of Sudan. Theor Appl Climatol 145:1113–1125. https://doi.org/10.1007/s00704-021-03690-1

Osman SOM, Saad ASI, Tadano S, Takeda Y, Konaka T, Yamasaki Y, Tahir ISA, Hisashi T, Akashi K (2022a) Chemical fingerprinting of heat stress responses in the leaves of common wheat by Fourier transform infrared spectroscopy. Int J Mol Sci 23. https://doi.org/10.3390/ijms23052842

Osman SOM, Saad ASI, Tadano S, Takeda Y, Yamasaki Y, Tahir ISA, Tsujimoto H, Akashi K (2022b) Probing differential metabolome responses among wheat genotypes to heat stress using Fourier transform infrared-based chemical fingerprinting. Agriculture (Switzerland) 12. https://doi.org/10.3390/agriculture12060753

Tahir ISA, Elbashier EME, Ibrahim MAS, Saad ASI, Abdalla OS (2020a) Genetic gain in wheat grain yield and nitrogen use efficiency at different nitrogen levels in an irrigated hot environment. Int J Agron 2020. https://doi.org/10.1155/2020/9024671

Tahir ISA, Mustafa HM, Idris AAM, Elhashimi AMA, Hassan MK, Fadul EM, Kurmut AMA, Eltayeb SM, Meheesi S, Hassan AO et al (2020b) Enhancing wheat production and food security in Sudan through scaling up improved technologies using innovation platforms. Int J Agric Sustain:376–388. https://doi.org/10.1080/14735903.2020.1787639

Tsujimoto H, Sohail Q, Matsuoka Y (2015) Broadening the genetic diversity of common and durum wheat for abiotic stress tolerance breeding. In: Advances in wheat genetics: from genome to field, pp 233–238. https://doi.org/10.1007/978-4-431-55675-6_25

6 Establishing Nitrogen-Efficient Wheat Production Systems in the Indo-Gangetic Plains Through the Deployment of BNI Technology

Satoshi Tobita, Tadashi Yoshihashi, Masahiro Kishii, Guntur Ventaka Subbarao, and Arun Kumar Joshi

Abstract

Because of the rapid nitrification process in soil, more than half of the applied nitrogen fertilizer is not utilized by crops and is lost to the environment, which resulted in low nitrogen-use efficiency (NUE) in agricultural production systems. Suppression of nitrification is key to improving NUE, allowing for more N to stay in the soil and to be used by crops. Biological nitrification inhibition (BNI) is a function of plants that suppresses the nitrification process using special chemical compounds—biological nitrification inhibitors (BNIs)—from the roots, which block key enzymes involved in the nitrification process. For wheat, one of the most fertilized crops in the world, its wild relative *Leymus racemosus* was found to have a high BNI capacity. The trait was successfully transferred into international wheat varieties, such as Munal and Roelf. The BNI-enabled varieties/lines have been proven to suppress nitrification and utilize soil nitrogen more efficiently than the original varieties. The SATREPS project, "Establishment of nitrogen-efficient wheat production systems in Indo-Gangetic Plains by the deployment of BNI-technology," was started in 2022 for the social implementation of these BNI-enabled wheat lines in the Indo-Gangetic Plains (IGPs), a food basket for 1.4 billion people, where farmers apply extremely high rates of N fertilizer to wheat. The project aims at proposing of the new nitrogen-efficient wheat production systems using BNI technology and demonstrating in farmers field that no decrease in grain production nor quality under 30% less of N fertilizer application with BNI technology. As the deployment of BNI technology in the IGPs will be achieved by having an Indian elite wheat variety with enhanced BNI capacity. The project advances breeding efforts to transfer the BNI ability from BNI-enabled international wheat varieties to Indian elite varieties. A molecular marker of the alien chromosome segment (Lr#n-SA), conferring

S. Tobita (✉)
College of Bioresources Science, Nihon University, Fujisawa, Kanagawa, Japan
e-mail: tobita.satoshi@nihon-u.ac.jp

T. Yoshihashi · M. Kishii · G. V. Subbarao
Japan International Research Center for Agricultural Sciences (JIRCAS), Tsukuba, Ibaraki, Japan
e-mail: yoshihashit0216@jircas.go.jp; kishiim0329@jircas.go.jp; subbarao@jircas.affrc.go.jp

A. K. Joshi
Borlaug Institute for South Asia, Delhi Liaison Office, New Delhi, India
e-mail: a.k.joshi@cgiar.org

T. Nagamine et al. (eds.), *Plant Genetic Resources (PGR) for Sustainable Crop Production*, Sustainable Development Goals Series, https://doi.org/10.1007/978-981-96-7117-5_6

the BNI traits, was developed in the current project, and the marker is used to select for enhanced BNI capacity. To demonstrate the environmental benefits of BNI traits in fields, the following methodologies were tried: (1) collecting N_2O gas emitted from the rhizosphere and (2) collecting soil water at different depths to investigate the dynamics of N, especially NO_3-N, in the soil. After verification, the two methods will be applied in experimental fields in the IGPs to monitor and compare the emission of N_2O and the emergence of NO_3-N in BNI-enabled field with non-BNI field.

Keywords

Biological nitrification inhibition (BNI) · Fertilizer-N · Indo-Gangetic Plains · NUE · Wheat

6.1 Nitrogen in the Current Environment

6.1.1 Industrial Production of Ammonia by the Haber-Bosch Process

Nitrogen is one of the most essential elements for all biological systems, constituting proteins, nucleic acids, pigments, and neurotransmitters. Nitrogen is abundant and ubiquitous in the Earth's troposphere, and most of it is in the form of N_2 ($N \equiv N$), where two N atoms are strongly triple-bonded. Because N_2 gas is stable and unreactive, living organisms cannot directly utilize N in the air. Biologically accessible N is categorized as reactive nitrogen (Nr), including all N forms other than N_2 gas, such as inorganic NH_3 and NO_x and organic N. Under natural conditions, Nr is converted from unreactive N_2 through biological nitrogen fixation (BNF) to produce ammonium-N by symbiotic microorganisms in leguminous plants or through atmospheric deposition to produce oxides of N (NO_x) via high-energy events such as lightning and wildfire. All organisms depend on the scarce Nr available on Earth, including that in agricultural systems. The nitrogen sources used to nourish the crops were only green manure, animal/human excreta, guanos, or saltpeter. Therefore, agricultural productivity stagnated until humans acquired the method for mass ammonia production in the early nineteenth century. This method is called the Haber-Bosch process (named after the two inventers, Fritz Haber and Carl Bosch) and is used to synthesize ammonia (NH_3) on an industrial scale using nitrogen from the air. The process proceeds under high temperatures and pressure to break the triple bond; therefore, this process requires a large amount of energy, equivalent to 1% of worldwide energy consumption (Capdevila-Cortada 2019).

Since the Haber-Bosch process has been implemented, producing as much ammonia as desired has become possible. Ammonia is a versatile source of chemical fertilizers, explosives, and other N-containing chemical materials. Thus, the process dramatically increases the Nr inputs into anthropogenic activities and natural ecosystems. The annual global industrial production of ammonia was estimated to be 150 million metric tons of N (150 TgN) in 2021 (USGS 2023), 80% of which is used for making N fertilizer (mainly urea). Modern agriculture and food supply are highly dependent on this anthropogenic N.

6.1.2 Modern Agriculture Depends Heavily on Chemical Nitrogen Fertilizer

Historically, the agricultural benefits of Haber-Bosch nitrogen were realized as part of the Green Revolution, which first started occurring in the 1950s. High-yielding varieties of wheat and rice, bred by introducing semi-dwarfing genes, had higher and more stable yields without lodging in response to higher N fertilizer application compared to traditional varieties. These varieties were disseminated along with sufficient chemical N fertilizer; thus, the production of wheat and rice was significantly increased to avoid food crises in Asia and support the growing world population. Regarding wheat, Mexico, once an importer in the 1940s, achieved self-sufficiency in the 1950s and became an exporter of wheat in the 1960s. High-yielding wheat cultivars were introduced, along with irrigation systems, to the Indo-Gangetic Plains, the northern part of the Indian Subcontinent, currently a food basket for India, Pakistan, and Nepal. Notably, the Green Revolution was achieved not only through high-yielding varieties but also through increased accessibility of chemical N fertilizers to local farmers, attributed to the Haber-Bosch process. Figure 6.1 shows the causal relationship. World population growth is supported by increased food (crop and meat) production, or food production per unit area, as arable land has not increased over the past century. Food production is proportional to ammonia/N-fertilizer production via Haber-Bosch (Hayashi et al. 2021). With the Green Revolution, agricultural systems have changed in several ways. Diversified cropping systems with various crops, especially legumes, have declined, and crop cultivation has been separated from animal rearing. This is why crop production has accelerated to depend more on chemical N fertilizer. Erisman et al. (2008) estimated that approximately 50% of the human population is alive due to Haber-Bosch nitrogen production. They also mentioned the increasing influence of Haber-Bosch nitrogen on growing bioenergy and biofuel production.

6.1.3 Low NUE and Global N Cycle in Crisis

NUE is defined as the output/input ratio of N in an agricultural system. Lassaletta et al. (2014) calculated the NUE of 124 countries over the past 50 years using the Food and Agriculture Organization (FAO) database. It has been shown that NUE was 65% in the early 1960s, decreased to approximately 45% in the 1980s, and became stable at approximately 47% in recent years. Comparing NUE among crops, rice and wheat that received a relatively high rate of N-fertilizer had a lower NUE of approximately 40%, whereas soybean had a higher NUE of 80%. Figure 6.2 shows a quantitative estimation of the nitrogen budget in world agriculture using datasets from 2011 to 2015 (Zhang et al. 2021), resulting in nitrogen assimilation by crops of only 78 TgN/year. In contrast, nitrogen inputs to agricultural lands total 169 TgN/year, comprising synthetic N fertilizer (98 TgN), manure (22 TgN) applications, BNF (32 TgN), and atmospheric deposition (17 TgN). This means that more than half (54%) of the nitrogen applied to agricultural

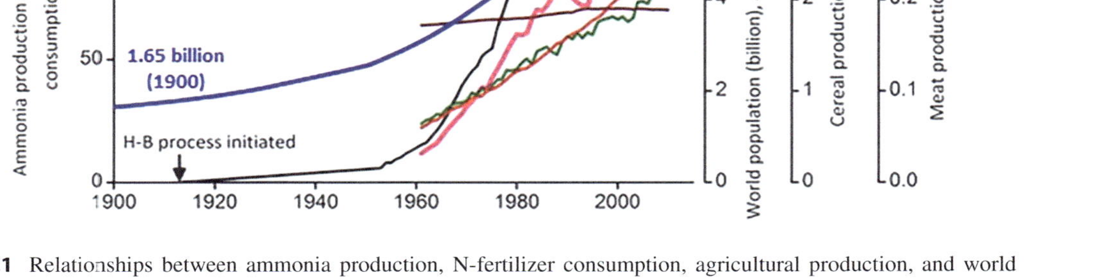

Fig. 6.1 Relationships between ammonia production, N-fertilizer consumption, agricultural production, and world population growth from 1900 to 2000 (Hayashi et al. 2021)

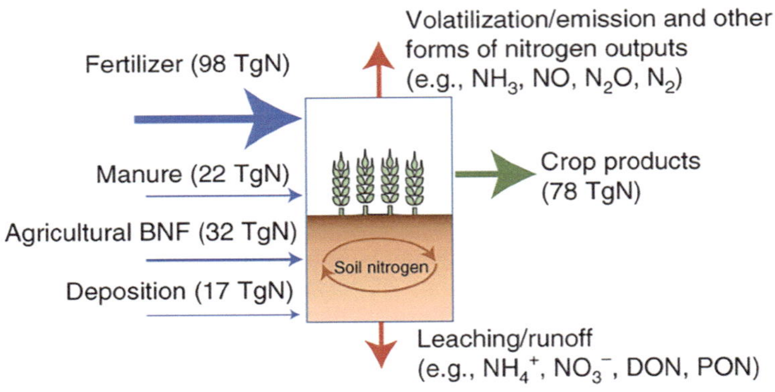

Fig. 6.2 Quantitative estimation of global N budget in agricultural production systems (Zhang et al. 2021)

lands was not used nor harvested by crops. Although only a small portion of this unused nitrogen was incorporated as soil organic nitrogen, most was lost to the surrounding environments. The loss of N from the root zone has significant economic consequences, accounting for US$17 billion annually (Subbarao et al. 2006a). Nitrogen is released into the air as volatilized NH_3, and emitted as NO, N_2O, and N_2, and into the hydrosphere as inorganic N (NH_4^+ and NO_3^-) and organic N (dissolved and particulate). It is noted that except for N_2 gas, Nr includes all other chemical N species discharged into the environment.

Erstwhile, the Earth's system has managed this amount of Nr through denitrification to maintain the global N cycle (returning it to stable N_2 gas) and through deposition into soil or the sea. However, in the modern world, anthropogenic processes produce a significant amount of additional Nr through industrial N fixation (i.e., the Haber-Bosch process for fertilizers), BNF, and the combustion of fossil fuels. When the Nr from both natural and anthropogenic sources is incorporated into the global nitrogen cycle, the Earth's system cannot properly maintain the appropriate quantity (or ratio) of global Nr. For this reason, Rockström et al. (2009) and Steffen et al. (2015) recognized that the global N cycle has already exceeded its planetary boundary (zone of uncertainty) and is at high risk, along with biodiversity loss, among the other nine Earth system processes (Fig. 6.3, updated by Richardson 2023).

6.2 BNI (Biological Nitrification Inhibition) as a Plant Function

6.2.1 The Urgent Need to Improve NUE in Agricultural Systems

Excess amounts of Nr species in the ecosystem have detrimental effects; for example, water quality is degraded by ammonium-N and nitrate-N as well as by dissolved or particulate organic N, whereas the air is polluted by NO and N_2O gases, which are potent greenhouse gases (GHGs) that accelerate climate change. Therefore, we must pay attention to the increasing nitrogen load on the Earth and implement measures to reduce such threats caused by Nr. More importantly, agriculture contributes to 75% of Nr production, which is a major factor in the high risk to the nitrogen cycle in sensitive ecosystems. The average value of NUE in cereal production worldwide has remained low at 33% for two decades (Raun and Johnson 1999). Low NUE is proof of this; therefore, improving NUE through agricultural technologies to mitigate the adverse effects of reactive nitrogen is a pressing issue. Inherently, NUE improvement leads to adequate N use to meet crop demands; thus, sustainable nitrogen use ensures food security for the increasing global population.

Several research opportunities exist for technological development to improve NUE, as

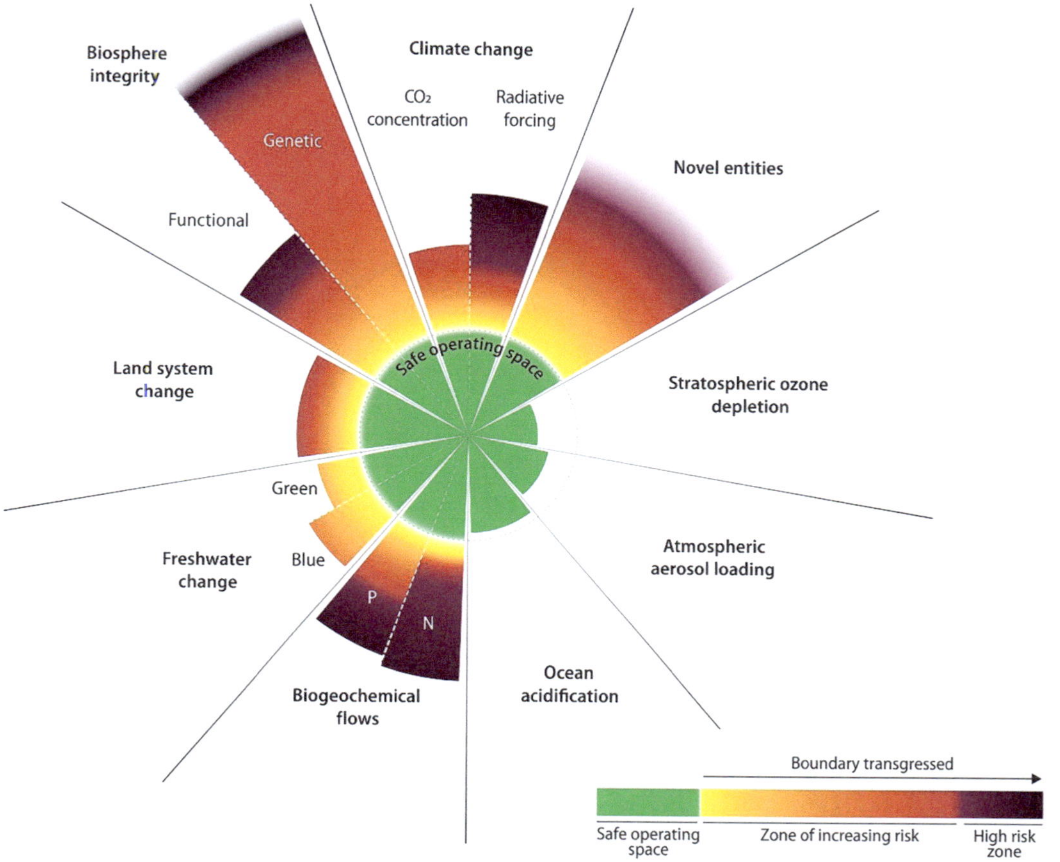

Fig. 6.3 Current status of nine earth system processes in planetary boundaries (Richardson et al. 2023)

reviewed by Udvardi et al. (2021). These include (i) improving the synchrony of soil N supply and crop demand both temporally and spatially (e.g., split N application); (ii) enhancing the efficiency of N fertilizers with novel formulations (e.g., polythene-coated and slow-release fertilizers) or containing effective inhibitors (urease and nitrification inhibitors); (iii) a mixture of inorganic-N fertilizers (chemical) and organic-N-based fertilizers (manure and crop residue); (iv) precision agriculture to use big data by monitoring crops and soil/environments (with drones or crop sensors); (v) BNF enhancement and incorporation of legumes; and (vi) crop breeding for higher NUE based on plant-microbe interactions.

6.2.2 Nitrification Pathway and Its Biochemical Mechanisms

For the improvement of NUE, it is important to comprehend the chemical transformations of N species in agricultural soils, because nitrogen lost to the environment emerges exclusively from the soil through the biochemical (enzymatic) activities of microorganisms underground, except for volatilized ammonia (NH_3), which accounts for 10–20% of the total N inputs (van Grinsven et al. 2015). Figure 6.4 shows the pathways of the chemical species of N in the soil, illustrating the transformation of ammonium-N (derived from chemical N fertilizer and other forms of nitrogen inputs, such as BNF, manures, and crop residues) to nitrate-N through the nitrification pathway

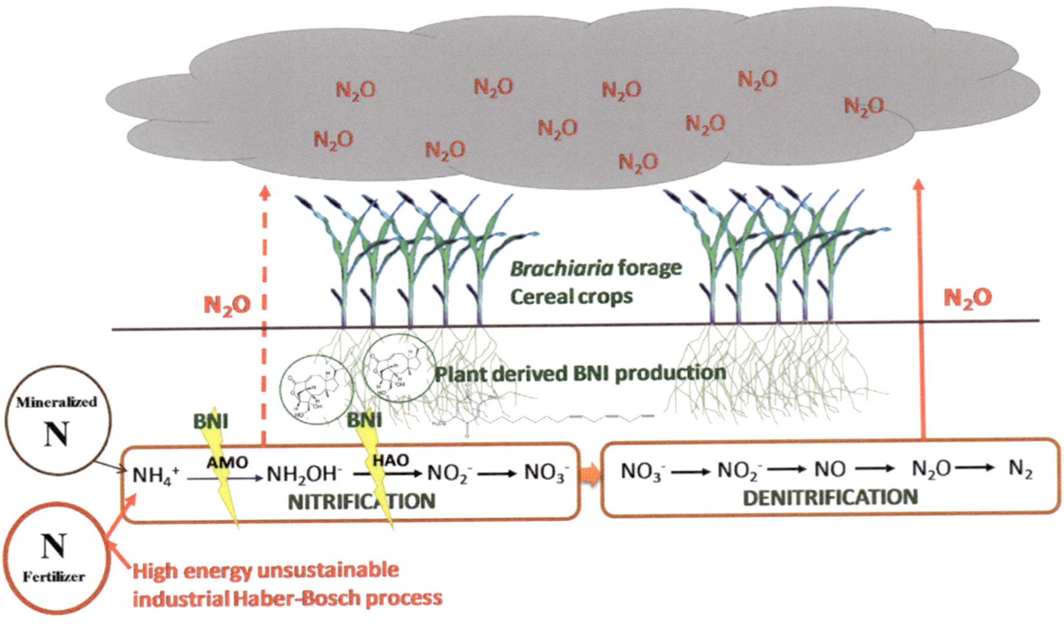

Fig. 6.4 Transformation of N chemical species in the soil through biochemical processes. Blocking sites of BNIs are also shown. (Courtesy of JIRCAS)

(left half). Nitrate-N is then reduced to form NO, N_2O, and N_2, through the denitrification pathway (right half). In modern agriculture, which depends heavily on synthetic N fertilizers, the nitrification rate from NH_4^+ to NO_3^- is estimated to be less than 10 days (Subbarao and Searchinger 2021). Since anion NO_3- does not generally bind to the soil surface, unlike cation NH_4^+, it quickly leaches with the flow of soil water into the groundwater. It enters the denitrification pathway, through which nitrogenous gases are generated. From a plant physiology perspective, such a situation is not favorable for the growth of annual crops because most of the N minerals (NH_4^+ and NO_3^-) would disappear around their roots at the highest demand unless the timing and place of N fertilizer application are considered. This is a direct explanation for the low NUE, and an improvement in NUE can be achieved if we can keep NH_4^+ in the rhizosphere for a more extended period by decelerating the nitrification process.

NH_4-oxidizing reactions in the nitrification pathway are governed by three groups of soil microorganisms: (1) ammonia-oxidizing bacteria (AOB), (2) ammonia-oxidizing archaea (AOA), and (3) nitrite-oxidizing bacteria (NOB). The first two groups are responsible for the transformation of NH_3 to NH_2OH and of NH_2OH to NO_2-, catalyzed by the corresponding enzymes ammonia monooxygenase (AMO) and hydroxylamine oxidoreductase (HAO), respectively. NOB participates in the transformation of NO_2^- to NO_3^- with NXR (nitrite oxidoreductase [NOR]), as in Fig. 6.4. The inhibition of these microbial or enzymatic activities results in a reduced nitrification rate.

Synthetic chemicals such as allylthiourea (AT), nitrapyrin (NP), dicyandiamide (DCD), and 3,4-dimethylpyrazole phosphate (DMPP) can inhibit nitrification. Although their usage and recommended doses differ, these inhibitors target the AMO, the first step in nitrification. As summarized by Ayiti and Babalola (2022), these synthetic nitrification inhibitors have several shortcomings, such as difficulties in application, high cost, environmental pollution, food safety, transient (easily washed out with water), and discriminatory effectiveness, which are significant disadvantages compared with BNIs, as written in the following subsections.

6.2.3 Discovery of the BNI Phenomenon in Ecosystems of the Tropics

Reportedly, in 1983, scientists from the International Center for Tropical Agriculture (CIAT) found that in Colombia and Brazil, fields of *Brachiaria humidicola*, a tropical grass species also known as creeping signalgrass, generally had low levels of soil NO_3-N. This was the first report on the BNI phenomenon (Ishikawa et al. 2003). They conducted comparative studies with other tropical forage grasses and indicated the possibility of nitrification inhibition in the soil by *B. humidicola*. In addition, they observed that the grass species was less responsive to the application of inorganic fertilizer-N, contrary to other grass species. The team from the Japan International Research Center for Agricultural Sciences (JIRCAS), led by Dr. Guntur Venkata Subbarao, eventually succeeded in establishing BNI technology. They showed that the exudates of *B. humidicola* roots suppressed nitrification (delayed NO_3^- generation) in soil incubation experiments because of a decreased AOB population, as estimated using the MPN (most probable number) method.

Consistent with the first report on the BNI phenomenon in Latin American tropical grasslands (Ishikawa et al. 2003), a French group also reported the possible allelopathic inhibition of nitrification by root exudates from the grass *Hyparrhenia diplandra* (Hack.) Stapf in a wet tropical savanna in Côte d'Ivoire, West Africa (Lata et al. 2004). From an ecological perspective, they indicated that the nitrification process plays a key role in the functioning of savanna natural ecosystems for plant N acquisition and N loss, and nitrification remains low, presumably owing to the nitrification inhibition ability of several savanna plant species, an adaptive trait to thrive in N-limiting environments (Lata et al. 2004). These findings concluded that BNI might be a widespread phenomenon observed in various plant species in diversified ecosystems. BNI is also thought to be a trait that presumably evolved to increase plant BNI activity (i.e., BNI capacity) in N-limiting environments.

6.2.4 Assay for the Determination of BNI Activity

The quantitative evaluation of the nitrification rate is critically essential for BNI studies. Subbarao et al. (2006b) established a bioluminescence assay using recombinant *Nitrosomonas europaea* to detect and quantify the activity of nitrification inhibition in plant-soil systems, as expressed in AT Units or ATU, where 1 ATU was defined as the inhibition of NO_2^- production caused by 0.22 μM of allylthiourea (AT), a synthetic nitrification inhibitor. This novel protocol confirmed that 20 ATU of BNI activity released from *B. humidicola* roots inhibited NO_3^- formation during a 55-day incubation and remained functionally stable in the soil for 50 days. Both the AMO and HAO enzymatic pathways in *Nitrosomonas* were effectively blocked by BNI activity. Therefore, this assay has been used as an effective method to characterize and determine the BNI activity in exudates, extracts, or homogenates of plant roots.

BNI influences the biochemical nitrification process performed by nitrifiers in the soil, such as AOB and AOA. Therefore, the existence of BNI should prove how the plant suppresses the populations of AOB and AOA in the rhizosphere, as measured by the expression of *amoA* gene. Moreover, BNI is a phenomenon based on a mechanism where certain compounds from a plant, such as biological nitrification inhibitors, affect AMO or HAO enzymes (Fig. 6.4). Thus, studies on isolation, concentration, characterization, structure determination, and ultimately identification of the compounds (biological nitrification inhibitors) are an effective way to obtain physical evidence of BNI. For example, brachialactone, a diterpenoid, was found as a biological nitrification inhibitor from *B. humidicola*, blocking AMO but not HAO (Subbarao et al. 2009).

6.2.5 Screening Food and Forage Crops for High BNI Ability

Standardizing the bioluminescent assay for detecting and quantifying BNI activity (expressed as ATU), the BNI research team at JIRCAS has intensively promoted the screening of plant species or cultivars/varieties within a specific crop. In collaboration with CIAT, they tested several tropical food and forage crops (Subbarao et al. 2007a). However, at that moment, neither of the world's three major crops (wheat, maize, and rice) nor barley showed nitrification inhibition.

In a successful collaboration with the International Crops Research Institute for the Semi-Arid Tropics (ICRISAT), the JIRCAS team recognized a wide range of BNI capacity within accessions of sorghum (*Sorghum bicolor* L.), the fifth most important crop. Among those, accessions from Sub-Saharan region had a relatively higher BNI capacity. For sorghum, sorgoleone, an already recognized allelopathic compound to suppress weeds, was isolated from the hydrophobic fraction of the root exudates with significant nitrification inhibition (Subbarao et al. 2013). Sarr et al. (2020) reported that releasing higher amounts of sorgoleone causes alterations in the soil microbial community and has great potential to inhibit the abundance of AOA (but not AOB) and soil nitrification.

BNI-enabled maize or corn (*Zea mays* L.) has long been awaited because the crop is cultivated in most of the world in terms of yield and receives more than 20% of the chemically produced N fertilizers. Since Dr. Tadashi Yoshihashi, a chemist of JIRCAS, had found BNI activity in root exudates from a commercial corn variety in 2017, intensive studies have been conducted on the BNI activity of maize over a short period. Otaka et al. (2022, 2023) reported that two hydrophobic BNI compounds were isolated and identified from the root exudates, one of which was a new naphthoquinone named "zeanone." The other was 6-methoxy-2(3H)-benzoxazolone (MBOA), which showed potent nitrification inhibition (ED_{50} = 2 μM and 0.67 μM, respectively, indicating that these doses suppress half of the nitrification activity estimated by ATU).

6.2.6 Production of BNI-Enabled Wheat Lines through Pre-breeding

As the second-largest crop in terms of production after maize, wheat (*Triticum aestivum* L.) has recently required more than 18% of the world's total nitrogen fertilizer, the highest percentage among all crops until 2014 (IFA 2022). Therefore, improving the NUE in wheat production systems is a global challenge to reduce N leakage into the environment. To address this issue, scientists from JIRCAS and the International Maize and Wheat Improvement Center (CIMMYT), based in El Batán, Mexico, have evaluated many wheat varieties with higher BNI ability. Unfortunately, wheat genetic resources have no significant BNI capacity (Subbarao et al. 2007a). As part of the exploration of alien genes from wild relative species for wheat improvement, a group of Japanese wheat breeders/geneticists (Dr. Hisashi Tsujimoto and Dr. Masahiro Kishii) found that a wild relative species of wheat, *Leymus racemosus* L. (Photo 6.1), had high BNI capacity, which effectively suppressed soil nitrification (Subbarao et al. 2007b). For the BNI capacity of *L. racemosus* to be transferred into cultivated wheat without any undesirable effects derived from alien genes, considerable efforts have been made to select interspecific hybrid lines with the minimum chromosomal segments of *Leymus* for a higher BNI capacity. Eventually, a genetic stock of wheat with the *Leymus* chromosome segment (Lr#n-SA) showing high BNI capacity was obtained under the background of "Chinese Spring," an old wheat cultivar often used as a genetic and grain quality standard. Then, the superior genetic stock was crossed with international wheat cultivars "Munal" and "Roelf" to successfully produce BNI-enabled BNI-Munal and BNI-Roelf (Subbarao et al. 2022). This indicated that the genetic barrier between the crop and its wild relatives was successfully removed through pre-breeding. It is still worth exploring existing wheat landraces or relatives for their

Photo 6.1 Natural population of *Leymus racemosus* in Bulgaria (©Bulgarian Flora Online)

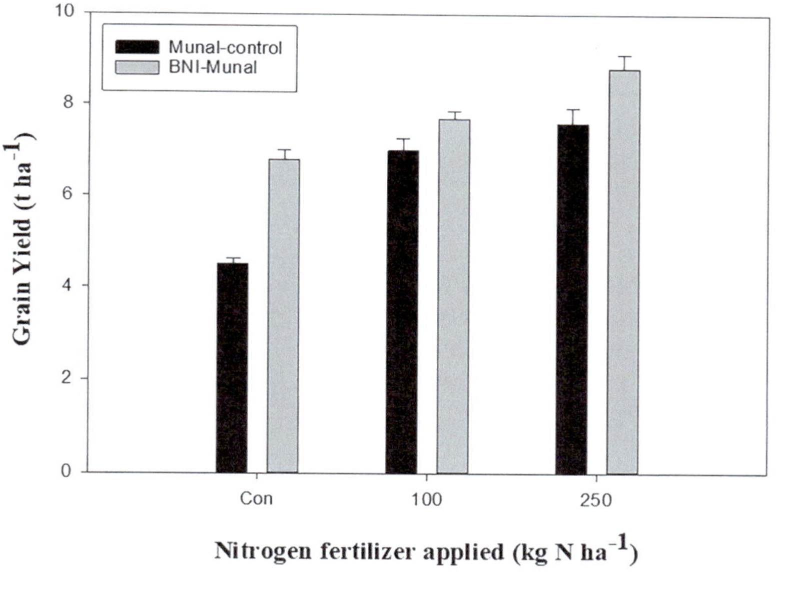

Fig. 6.5 Yield response of wheat variety Munal and BNI-Munal line to different amounts of N application in field conditions of Japan. (Courtesy of JIRCAS)

BNI capacity as a genetic source for improving NUE in elite wheat varieties.

Soil nitrification activities, especially by AOA, were suppressed, and N_2O emission was reduced in a laboratory incubation test with the soil collected from the root zone of field-grown BNI-Munal. In an experimental field in Japan, the grain yield of BNI-Munal was superior to "Munal" under different nitrogen fertilizer applications, including no N input, showing that NUE was improved by the "acquired" BNI capacity (Fig. 6.5 and Photo 6.2). Reducing N fertilizer applications to gain the same grain yields might be possible. It was shown that BNI-Munal absorbed more N than "Munal" in the field experiment and grain quality was not negatively affected in breadmaking properties (Subbarao et al. 2022). Therefore, BNI-Munal and other BNI-enabled international wheat lines (e.g., BNI-Roelf and BNI-Vorobey) themselves can be disseminated to wheat fields as a nitrogen-efficient wheat cultivar upon adaptation and can

Photo 6.2 Observation of wheat variety Munal (left) and BNI-Munal line (right) plants under no N application at maturity in a 2020/2021 experiment in Japan

be used as a breeding material for nitrogen-efficient elite wheat cultivars in particular environments and economical requirements. From the above, the BNI technology with BNI-enabled crops is expected to be a possible solution for the global N crisis.

6.3 Deployment of BNI Technology in Wheat Production Systems in the Indo-Gangetic Plains (IGPs)

6.3.1 Rationale for Introducing BNI Technology in the IGPs

The "proof of concept" for BNI technology has been established and shown to work in the field during the 2019/2020 and 2020/2021 seasons (Subbarao et al. 2022). Thus, it is now ready for large-scale testing to determine its suitability for deployment in major wheat-growing regions of the world with high nitrogen application. Ultimately, the most suitable site was selected to implement a project aimed at deploying BNI technology in the Indo-Gangetic Plains, an area that benefited most from the Green Revolution and now serves as the breadbasket feeding 1.4 billion people in India. India is the world's second-largest wheat producer (Fig. 6.6).

As mentioned earlier, wheat is among the most heavily fertilized crops. According to Sapkota et al. (2020), in India, wheat receives about 120 kg N ha^{-1} (on average) but can reach up to 300 kg N ha^{-1} (in high-N input systems where yield potential is close to 9 t ha^{-1}, in areas such as Punjab and Haryana), while it is medium to low (70–120 kg N ha^{-1}) as we move towards the eastern parts of the IGPs (Praveen et al. 2020). The N application rates in this region are incredibly high, far exceeding the crop demand (Bora 2022). Therefore, there is sufficient rationale to introduce BNI technology in this area to improve its low NUE (<30%) and reduce the amount of N fertilizer. Achieving this agronomic target, we can expect benefits in two aspects. One is the economic benefits for farmers, such as reducing the cost and labor for fertilization on their farms. Also, the Central Government of India has a tangible policy to reduce the consumption and import of N fertilizers, as farmers tend to use too much urea (NITI Aayog 2015), and wants to save the budget for subsidies to fertilizer users, which supports the project politically. Another benefit is environmental; locally, it alleviates nitrate pollution in aquatic environments, and globally, it mitigates the emission of N_2O gas from fields cultivated with BNI-enabled wheat varieties.

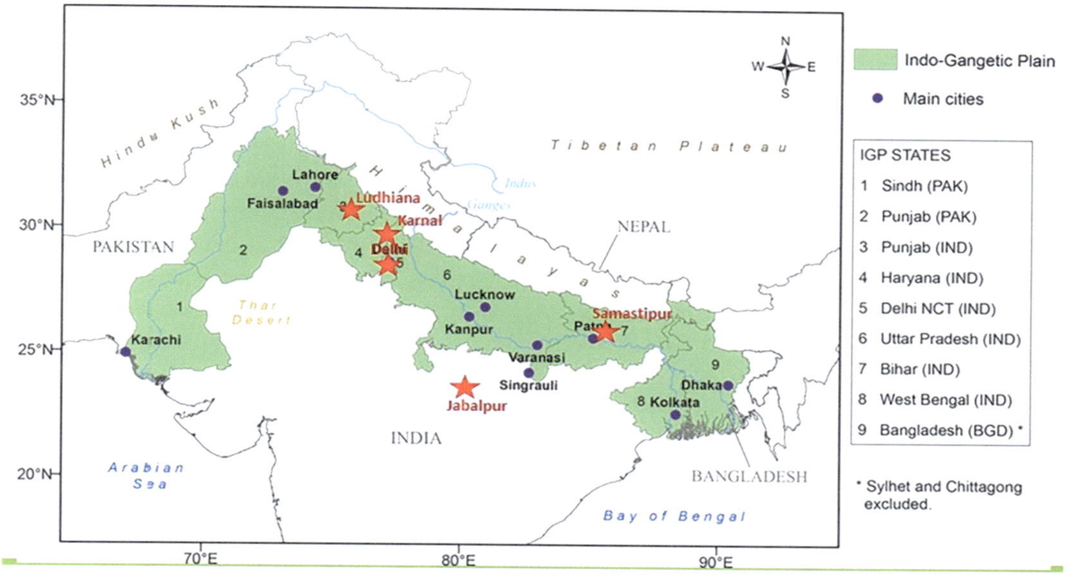

Fig. 6.6 Map of the Indo-Gangetic Plains (IGPs) expanding from Pakistan, India, and Bangladesh, with administrative states numerically indicated, as well as main cities (●) and project sites (★). (Modified by the authors from the original, copyrighted by the European Geosciences Union through the journal *Atmospheric Chemistry and Physics*)

6.3.2 The SATREPS Project for the Deployment of BNI Technology

For the above reasons, scientists from JIRCAS and the Borlaug Institute for South Asia (BISA) proposed a collaboration project with institutions in Japan and India for the social implementation of BNI technology in the IGPs. BISA is an international, regional research institute established by the Government of India in collaboration with CIMMYT and the Indian Council of Agricultural Research (ICAR). The SATREPS project titled, "The project for the establishment of nitrogen-efficient wheat production systems in Indo-Gangetic Plains by the deployment of BNI-technology," was approved in 2021, commenced in 2022, and will continue until 2027. BISA and its Indian partners received funding from JICA, whereas the Japanese colleagues received support from JST.

During the preliminary phase of the project, both sides elaborated on the project outline and described it in a project design matrix (PDM). The project is led by JIRCAS on the Japanese side, with Tottori University and NARO (National Agricultural Research Organization under MAFF) participating as partners. On the Indian side, BISA serves as the project leader, with three ICAR (Indian Council of Agricultural Research) institutes—IIWBR (Indian Institute of Wheat and Barley Research), CSSRI (Central Soil Salinity Research Institute), and IARI (Indian Agricultural Research Institute)—as project partners. The project has several experimental sites in the IGPs, including Ludhiana (Punjab), Jabalpur (Madhya Pradesh), and Samastipur (Bihar), which are HQs and stations for BISA, as well as Karnal (Haryana), the base for the Indian Institute for Wheat and Barley Research (IIWBR) and the Central Soil Salinity Research Institute (CSSRI), and Delhi, the base for the Indian Agricultural Research Institute (IARI), also known as Pusa (Fig. 6.6).

The project aims to propose new nitrogen-efficient wheat production systems using BNI technology at the project sites, where the following numeric indicators will be demonstrated: (1)

the same level of grain yield and quality with a 30% reduction in N application, (2) a 10–20% reduction in N leaching, and (3) a 20% reduction in N_2O emission at the project sites. Of course, the deployment of BNI technology in the IGPs will be achieved by having an Indian elite wheat variety with enhanced BNI capacity. Since breeding and registering a new variety may take more than 5 years, the term of the project is put in the upper goal, as well as the recognition of the benefits of BNI technology by all stakeholders involved in wheat production.

6.3.3 Activities and Progress in Research Subjects of the SATREPS Project

The project has four research subjects to proceed with the social implementation of BNI technology in the IGPs.

The first subject is to transfer the BNI ability into Indian elite wheat varieties recommended in different areas of the IGP through crossing and generation advancement. A breeding strategy is in place to incorporate the BNI trait into the background of the best wheat varieties (Mishra et al. 2024). The varieties nominated to be crossed as recipients are DBW303, DBW222, and DBW327 for the North Western Plain Zone (NWPZ), DBW252 and HI1612 for the North Eastern Plain Zone (NEPZ), DBW187 for NWPZ and NEPZ, and DBW110 for the Central Zone (CZ). The donors are BNI-Munal, BNI-Roelf, and BNI-Vorobey. Crosses were made in Ludhiana and Jabalpur by BISA, Karnal by IIWBR, and Delhi-Pusa by IARI during the 2021/2022 rabi season, and the resulting F_1 and BC_2F_2 were transferred to generation-advancing facilities in Dalang Maidan, Himachal Pradesh, during the off-seasons of 2022, 2023, and 2024, respectively. In the *rabi* season of 2023/2024, the generation reached different stages of the breeding program depending on the number of backcrosses. For instance, crosses reached F_5, BC_1F_4, or BC_2F_3 to be selected for BNI traits, and the lines are going to be checked for favorable agronomic characteristics, e.g., high and stable yield and resistance to rusts.

A simple method was established for selecting progenies with the chromosome segment (Lr#n-SA) conferring BNI traits, using molecular markers, which was verified by in situ microscopic observation (GISH) in this project. Within the term of the SATREPS project, the BNI-enabled lines with a background of Indian elite wheat varieties will be submitted for the official variety release process by the Government of India.

Owing to the advantage of having multiple experimental sites with different social and environmental conditions in the IGPs, the project is implementing a multi-location linkage trial in Ludhiana, Karnal, Jabalpur, and Samastipur (Fig. 6.6). The same set of wheat varieties/lines ("Munal," "Roelf," "Vorobey," and their BNI-enabled lines) and several local check varieties were cultivated to observe their agronomic performance under different amounts of N application. The same split-plot design was applied to all these experimental sites with four levels of N treatments: recommended (100%), 75%, 50%, and no application (0%). It is noted that due to varying farming practices across different zones of India, the recommended N application rate differs by site: 150 kg ha^{-1} in Delhi, Karnal, and Ludhiana, 130 kg ha^{-1} in Samastipur, and 120 kg ha^{-1} in Jabalpur. Field observations and preliminary analyses at each site showed that the effect of BNI ability in terms of growth and yield is expressed more obviously in N-limited conditions at all sites (Photo 6.3). Agronomic results of the multi-location experiment will be published in the future.

The N nutritional status of plants and soils in the multi-location trial, most importantly NUE and nitrification rate, is being analyzed. Climate data are monitored by weather stations equipped in each experimental site. So, G x E x M analysis is planned with these data to elucidate the effect of environment and management, separately or in interaction, on the expression of BNI-related traits.

Secondly, the project aims to identify BNI substance(s) of the BNI-enabled wheat lines or

Photo 6.3 Multi-location linkage trial at Samastipur, India, highlighted the different performance between Munal and BNI-Munal under N-limited (zero applied) conditions (2023/2024 season)

Leymus racemosus, the original genetic source of BNI for wheat. The biochemical evidence of how these substances inhibit the nitrification process in the soil has been eagerly awaited, as BNI-enabled wheat is currently the most advanced crop among all BNI crops for social implementation. In the early years, attempts by the group of phytochemists in JIRCAS have been made to identify possible BNI substances from *L. racemosus* itself, but not yet successful because it is difficult to obtain enough quantity of exudates or extracts from the wild species with too slow growth habitat in Japan. On the contrary, the work to extract the BNI activity is going well, with samples from BNI-enabled wheat lines performing equally to wheat cultivars. Currently, an aliquot is applied to HPLC and ion-exchange chromatography to identify the BNI substance(s).

The demonstration of the environmental benefits of BNI technology in the field constitutes the third research subject of the project. Usually, it is challenging to measure flushes of N_2O from the soil using static chambers because the quantity of the emitted gas is too small relative to the large volume of the chambers. Hence, the project has been elaborated to develop a novel apparatus for collecting gas from the rhizosphere through an embedded PVC tube (Photos 6.4). So far, in Japan, smaller and/or more delayed emission of N_2O was observed from the soil planted with a BNI-enabled wheat line compared to the corresponding parental variety. The BNI trait is also expected to reduce the amount of nitrate-N in the rhizosphere and the loss of N from the agricultural fields. Currently, the project is accumulating data on the temporal and spatial distribution of NO_3-N in the experimental field of Japan, by quantifying nitrate in soil water collected by porous ceramic cups embedded in different soil depths. From these findings, the project will draw a picture of quantitative nitrogen dynamics (nitrogen chemical species in the soil-plant-atmosphere continuum) in wheat production systems influenced by BNI technology.

The fourth research subject comprises several activities aimed at promoting understanding of BNI technology among all stakeholders engaged in wheat production and milling, flour distribution, and sales, as well as policymakers on agriculture and food in India. BNI technology is a new plant-based technology; thus, it is necessary to recognize and describe its possible benefits and limitations by ex ante impact analysis. Moreover, once recognized by the stakeholders, BNI technology is thought to be disseminated very smoothly and quickly, as wheat farmers are only required to replace their variety with a BNI-enabled one.

Previously, an economist from JIRCAS, in collaboration with CIMMYT, conducted an ex

Photos 6.4 Installation of an apparatus for gas sampling from the rhizosphere (left) and a mouth of the sampling apparatus (right). (Courtesy of JIRCAS)

ante LCA of BNI wheat technology. They used scenarios based on the evidence of nitrification inhibition and N_2O emission reduction by BNI-enabled crops. They found that BNI-enabled wheat plants could effectively inhibit nitrification by 40% by 2050 and that the release of BNIs by roots occurs in soils with a pH of 5.5–7.0 at a depth of 30 cm. It was shown that when nitrification was suppressed by 40%, N fertilization and LC-GHG emission (including GHG emission associated with fertilizer production) would be reduced by 15.0 and 15.9%, respectively, and NUE would be improved by 16.7% at the field level (Leon et al. 2021).

In this line, as mentioned earlier, the project is running multiple site trials in different soils, including high pH, and climatic conditions in the IGPs in every *rabi* season. It is thought that the expression of BNI trait could be limited in alkaline soils, and there has long been an awareness that it is likely to limit the dissemination of BNI to all wheat-growing regions. A map of suitable regions influenced by BNI technology will be revised with real data on the performance of BNI wheat from the multi-location trial. Further, once BNI varieties are in the advanced stage of testing, a pre-release seed multiplication approach, as followed for UG99-resistant wheat varieties, would be extremely useful (Joshi et al. 2011).

6.4 Conclusion

Nitrogen plays a crucial yet complex role for humans and other living organisms. While it provides essential nutrients (Nr) for life, excess nitrogen (also Nr) can harm the environment. To ensure a sustainable future, it is critical to manage nitrogen effectively. Boosting food production while minimizing environmental harm is a significant challenge in agriculture today. Modern agriculture relies heavily on nitrogen-based chemical fertilizers, yet much of this nitrogen goes unused and is released into the environment. Improving NUE is, therefore, a holistic and justified solution to these challenges. Enhancing NUE aligns with the United Nations Sustainable Development Goals (SDGs) for 2030, specifically the goals of "Zero Hunger" (Goal 2), "Climate Action" (Goal 13), and "Life Below Water" (Goal 14). BNI technology offers an effective and practical way to improve NUE in agriculture, supporting SDGs 2, 13, and 14. In addition, BNI technology contributes to the 12th goal, "Responsible Consumption and Production," because it promotes the sustainable use of chemical fertilizer, and the 15th goal, "Life on Land," because the technology is based on the function of richly diversified terrestrial plants.

When BNI technology is implemented socially through crop varieties with enhanced

BNI capabilities, it shows great promise. BNI wheat is currently leading among BNI technologies, alongside BNI pasture systems in Latin America and New Zealand. Several BNI-enabled wheat lines with international wheat backgrounds ("Munal," "Roelf," "Vorobey," and recently "Borlaug 100") are already available. These lines are genetically stable and ready for distribution to wheat-producing regions of Nepal, a part of the IGPs. These BNI-enabled international wheat lines are being used as pre-breeding material to boost the BNI capability of local wheat varieties in India, through the current SATREPS project, as well as in Japan. It is noted that a major initiative is starting to promote BNI ability in all wheat worldwide, coordinated by CIMMYT and funded by Novo Nordisk.

Nevertheless, more clarity on the molecular mechanisms of nitrification inhibition in BNI-enabled wheat and additional evidence of benefits from farms and markets are needed. The SATREPS project is expected to fulfill these requirements.

References

Ayiti OE, Babalola OO (2022) Factors influencing soil nitrification process and the effect on environment and health. Front Sust Food Syst 6:Article 821994

Bora K (2022) Spatial patterns of fertilizer use and imbalances: evidence from rice cultivation in India. Environ Chall 7. https://doi.org/10.1016/j.envc.2022.100452

Capdevila-Cortada M (2019) Electrifying the Haber–Bosch. Nat Catal 2:1055

Erisman JW et al (2008) How a century of ammonia synthesis changed the world. Nat Geosci 1:636–638

Hayashi K et al (2021) History of nitrogen discovery and its human use (in Japanese). In: Nitrogen and environmental science. Asakura Shoten, pp 10–13. ISBN:978-4-254-18057-2

IFA (2022) Fertilizer use by crop and country for the 2017–2018 period. International Fertilizer Association (IFA), Paris. https://www.ifastat.org/consumption/fertilizeruse-by-crop. Last browsed on 31 Dec 2024

Ishikawa T et al (2003) Suppression of nitrification and nitrous oxide emission by the tropical grass *Brachiaria humidicola*. Plant Soil 255:413–419

Joshi AK et al (2011) Delivering rust resistant wheat to farmers: a step towards increased food security. Euphytica 179:187–196

Lassaletta L et al (2014) 50 year trends in nitrogen use efficiency of world cropping systems; the relationship between yield and nitrogen input to cropland. Environ Res Lett 9:105011. https://doi.org/10.1088/1748-9326/9/10/105011

Lata J-C et al (2004) Grass populations control nitrification in Savanna soils. Funct Ecol 18:605–611

Leon A et al (2021) An ex-ante life cycle assessment of wheat with high biological nitrification inhibition capacity. Environ Sci Pollut Res 29:7153–7169

Mishra CN, Sharma S, Pawar SK et al (2024) Phenotyping and breeding strategies for incorporating BNI trait in wheat varieties. Plant Physiol Rep 29:808–822. https://doi.org/10.1007/s40502-024-00843-y

NITI Aayog (2015) Raising agricultural productivity and making farming remunerative for farmers. An occasional paper. Government of India, p 46

Otaka J et al (2022) Biological nitrification inhibition in maize—isolation and identification of hydrophobic inhibitors from root exudates. Biol Fertil Soils 58:251–264

Otaka J et al (2023) Isolation and characterization of the hydrophilic BNI compound, 6-methoxy-2(3H)-benzoxazolone (MBOA), from maize roots. Plant Soil 489:341–359

Praveen KV et al (2020) Advancing with fertilizers in Indian agriculture: trends, challenges, and research priorities. Agric Econ Res Rev 33:49–60

Raun WR, Johnson GV (1999) Improving nitrogen-use efficiency for cereal production. Agron J 91:357–363

Richardson K et al (2023) Earth beyond six of nine planetary boundaries. Sci Adv 9:eadh2458. https://doi.org/10.1126/sciadv.adh2458

Rockström J et al (2009) Planetary boundaries: exploring the safe operating space for humanity. Ecol Soc 14:32

Sapkota TB et al (2020) Identifying optimum rates of fertilizer nitrogen application to maximize economic return and minimize nitrous oxide emission from rice-wheat systems in the Indo-Gangetic Plains of India. Arch Agron Soil Sci 66(14):2039–2054

Sarr et al (2020) Sorgoleone release from sorghum roots shapes the composition of nitrifying populations, total bacteria, and archaea and determines the level of nitrification. Biol Fert Soils 56:145–166

Steffen W et al (2015) Planetary boundaries: guiding human development on a changing planet. Nature 347:1259855

Subbarao GV, Searchinger TD (2021) A "more ammonium solution" that mitigates nitrogen pollution, boosts crop yields. PNAS 118(22):e2107576118

Subbarao GV et al (2006a) Scope and strategies for regulation of nitrification in agricultural systems—challenges and opportunities. Crit Rev Plant Sci 25:1–33

Subbarao GV et al (2006b) A bioluminescence assay to detect nitrification inhibitors released from plant roots: a case study with *Brachiaria humidicola*. Plant Soil 288:101–112

Subbarao GV et al (2007a) Biological nitrification inhibition (BNI)—is it a widespread phenomenon? Plant Soil 294:5–18

Subbarao et al (2007b) Can biological nitrification inhibition (BNI) genes from perennial *Leymus racemosus*

(Triticeae) combat nitrification in wheat farming? Plant Soil 299:55–64
Subbarao GV et al (2009) Evidence for biological nitrification inhibition in *Brachiaria* pastures. PNAS 106(41):17302–17307
Subbarao GV et al (2013) Biological nitrification inhibition (BNI) activity in sorghum and its characterization. Plant Soil 366:243–259
Subbarao GV et al (2022) Enlisting wild grass genes to combat nitrification in wheat farming: a nature-based solution. PNAS 118(35):e2106595118
Udvardi M et al (2021) A research road map for responsible use of agricultural nitrogen. Front Sustain Food Syst 5:Article 660155
USGS (2023) Nitrogen (fixed)—ammonia, USGS mineral commodity summaries, January 2023. Last browsed on 3 Feb 2023
van Grinsven HJM et al (2015) Losses of ammonia and nitrate from agriculture and their effect on nitrogen recovery in the European Union and the United States between 1900 and 2050. J Environ Qual 44:356–367
Zhang X et al (2021) Quantification of global and national nitrogen budgets for crop production. Nat Food 2:529–540

7 Fighting Climate Change with Quinoa: A Highly Nutritious Crop that Can Withstand Harsh Environments

Yasunari Fujita, Yasuo Yasui, Yukari Nagatoshi, Keisuke Katsura, and Yuji Tokura

Abstract

Quinoa is not only highly nutritious, but it is also the only crop that can be grown in the southern Altiplano region of Bolivia (at an altitude of around 4,000 m), which is a high plateau in a harsh environment with saline soil, drought, and frost. However, quinoa production is currently threatened by frequent extreme weather events and soil erosion caused by the expansion of agricultural land. Therefore, this project aims to develop production technology for resilient quinoa based on the conservation and management technology of sustainable agroecosystems using indigenous biological resources. By disseminating this technology, we will contribute to stable food production and increased national income in Bolivia. Furthermore, our ultimate goal is to contribute to strengthening global food security by applying this technology, which has been cultivated in extremely adverse environments, to diverse agricultural settings, not only in other arid regions of the world threatened by desertification. In this way, the overall objective of this research project is to contribute to the realization of the United Nations International Year of Quinoa (2013) through efforts to achieve the Sustainable Development Goals (SDG 2.4, to ensure sustainable food production systems based on resilient agricultural production systems; SDG 2.5, to conserve genetic diversity, including that of related wild species; SDG 15.3, to combat desertification).

Keywords

Quinoa · Bolivia · Resilience · Sustainable agriculture · Altiplano · Soil erosion · Climate change

Y. Fujita (✉) · Y. Nagatoshi
Japan International Research Center for Agricultural Sciences (JIRCAS), Tsukuba, Ibaraki, Japan
e-mail: fujitay0434@jircas.go.jp; nagatoshiy0299@jircas.go.jp

Y. Yasui · K. Katsura
Graduate School of Agriculture, Kyoto University, Sakyo, Kyoto, Japan
e-mail: yasui.yasuo.2a@kyoto-u.ac.jp; katsura.keisuke.2a@kyoto-u.ac.jp

Y. Tokura
Field Center of Animal Science and Agriculture, Obihiro University of Agriculture and Veterinary Medicine, Obihiro, Hokkaido, Japan
e-mail: ytoukura@obihiro.ac.jp

7.1 Introduction

Drought and desertification are causing serious social problems around the world, such as food instability, water scarcity, and poverty, and the establishment of a sustainable agricultural production system in arid regions is one of the most

T. Nagamine et al. (eds.), *Plant Genetic Resources (PGR) for Sustainable Crop Production*, Sustainable Development Goals Series, https://doi.org/10.1007/978-981-96-7117-5_7

important global issues. Under these circumstances, quinoa (Fig. 7.1), a nutritious crop native to the Andean region of South America that can be grown in marginal environments, is expected to play an important role in ensuring global food security and nutrition and in eradicating hunger (Fig. 7.2). The United Nations declared 2013 the International Year of Quinoa, calling for increased awareness of the nutritional, economic, and cultural value of quinoa, as well as the promotion of policies aimed at its sustainable cultivation and use. However, even in Bolivia, one of its countries of origin, quinoa production is threatened by frequent extreme weather events and soil erosion caused by unplanned agricultural expansion.

In our SATREPS project, we will develop quinoa production technology that strengthens resilience based on sustainable agricultural ecosystem conservation and management technology by improving the conservation of quinoa genetic resources, developing breeding materials that strengthen resilience, and improving fallow land management and crop-livestock integration. We will also aim to strengthen the quinoa production base in Bolivia by disseminating the technology developed.

Our research group was one of the first in the world to establish many quinoa inbred lines, successfully decipher the draft genome sequence, demonstrate the overall picture of the correlation between genotype and phenotype of quinoa lines, and lead the development of gene function analysis and generation promotion methods. Based on advanced plant science research, we conduct interdisciplinary research that aims to be implemented in society and includes the fields of genomics and genetics, molecular biology, plant physiology, cultivation, crop science, natural product chemistry, meteorology, environmental

Fig. 7.1 Quinoa grown in the marginal environments around the Salar de Uyuni

Fig. 7.2 Quinoa (on the left) served at a hotel near the Salar de Uyuni. Quinoa is often added to soups, salads, and juices, but it can also be eaten as a substitute for rice. The meat on this plate is llama meat (on the right)

soil science, grassland science, animal husbandry, veterinary science, and agricultural economics. It is characterized by interdisciplinary research that includes genomics and genetics, molecular biology, plant physiology, cultivation, crop science, natural product chemistry, meteorology, environmental soil science, grassland science, animal husbandry, veterinary medicine, and agricultural economics, and it is oriented toward social implementation.

Our SATREPS project consists of the following four research themes: (1) the development of genetic resources and construction of quinoa genome breeding platform; (2) the development of breeding materials to increase quinoa's resilience; (3) the development of conservation and management techniques for sustainable quinoa-based agricultural ecosystems using native genetic resources; and (4) the establishment of technical information extension networks in the Bolivian Altiplano.

7.2 The Development of Genetic Resources and Construction of Quinoa Genome Breeding Platform

In this section, we conducted the following studies on the collection and preservation of genetic resources and the construction of breeding infrastructure using these resources (Sect. 7.2.1), as well as on the construction of nested association mapping (NAM) populations to accelerate gene identification (Sect. 7.2.2), achieving the following results.

7.2.1 The Collection and Preservation of Genetic Resources and the Establishment of Breeding Infrastructure Using These Resources

Our previous genetic and genomic studies have revealed that quinoa (*Chenopodium quinoa*) can be divided into two major phylogenetic groups: lowland and highland types. Within the highland types, two distinct genetic groups, the northern and southern types, have been identified (Mizuno et al. 2020). The southern highland type is cultivated around the Uyuni Salt Flats, where conditions are exceptionally harsh due to high salinity and an annual rainfall of only ~200 mm. These characteristics make southern highland quinoa an invaluable model for foundational research on environmental stress tolerance and an important genetic resource for developing resilient crops. Known locally as "Real," southern highland quinoa varieties are characterized by their large and white grains, which enhance their commercial value over the other two types. Thus, during 2022 and 2023, we conducted field collections along a 500 km transect, spanning from Lake Titicaca to the Uyuni Salt Flats in Bolivia, focusing on landraces of quinoa, including Real varieties, as our primary target. In addition, we also targeted ajara (*C. quinoa* ssp. *melanospermum*), which is known as a free-living form of quinoa (Wilson 1990). Ajara is commonly found as a weedy associate in cultivated quinoa fields and in ruderal areas such as roadsides and vacant lots, making it a classic example of a ruderal wild plant. Unlike cultivated quinoa, ajara exhibits seed shattering, which allows it to grow independently without human intervention. Its seeds are black or brown and smaller than those of cultivated quinoa (Fig. 7.3). In some regions, the grains are harvested in their immature state before shattering and used for consumption (Mujica and Jacobsen 2006). Beyond its use in food consumption, ajara's shorter growing period compared to cultivated quinoa could make it a potential donor for early-maturity traits in quinoa breeding programs. Given that ajara easily hybridizes with cultivated quinoa, we consider it a valuable genetic resource for quinoa breeding.

To date, we have collected seeds from 82 cultivated landraces of quinoa and 201 related wild lines (ajara and unidentified species) in Bolivia. We also confirmed that the seeds of 43 quinoa lines listed in the PROINPA catalog retained their germination capacity. In total, we have collected 326 lines for quinoa genetic resources so far. At the Viacha field, phenotype investigations such as flowering time, seed color, and seed shattering for these samples are being conducted in collabo-

Fig. 7.3 Cultivated quinoa seeds (the large white seeds on the right) and quinoa's wild relative, ajara seeds (the small black seeds on the left)

ration with PROINPA, along with seed propagation. We are also conducting genome-wide genotyping for these lines using the dpMIG-seq technique developed in Japan (Nishimura et al. 2024). This next-generation sequencing-based technique enables the easy acquisition of approximately 2,000 markers on the quinoa genome without requiring purified DNA or specialized enzymes, such as restriction enzymes and ligase, which are essential for the RAD-seq technique. Thus, phenotypic and genotypic analyses are progressing in the Andean highlands of Bolivia. By 2025, we aim to establish an integrated quinoa genetic resource information system (IQ-GRIS) that combines these phenotypic and genotypic data, enabling the rapid development of quinoa varieties adapted to regional characteristics.

We are also advancing a long-term seed preservation program for quinoa genetic resources. Previous studies have recommended storing quinoa seeds with a moisture content of around 5–8% at −18 °C. To achieve this, we have initiated seed drying using silica gel and 50 ml Falcon tubes. By 2025, we plan to vacuum-package these seeds and begin storage at −18 °C. Moving forward, the operation of IQ-GRIS and the preserved seeds will be conducted within Bolivia. IQ-GRIS and the conserved seeds would serve as a foundation for advancing international research collaborations and breeding programs to enhance quinoa cultivation globally.

7.2.2 Construction of NAM Populations to Accelerate Gene Identification

In conjunction with Research Theme 2, we have supported the development of nested association mapping (NAM) population (Yu et al. 2008). In NAM populations, recombinant inbred line (RIL) populations are generated for each cross combination of F_1 from a single common parent and multiple lines (i.e., NAM populations are a collection of RILs with a common parent). The lines used for crossing with the common parent are called founders because they are the original lines for each RIL. In the NAM population, the genomic regions of the founders are fragmented by the common parent genome, eliminating the population structure that the original founder population had. This makes it possible not only to perform correlation mapping with high accuracy but also to efficiently find promising breeding material from the progeny of each RIL. With these points in mind, the collaborative research team agreed to use the large-grain Mañiqueña line, a superior line developed at PROINPA, as the common parent of the NAM population and to use six founder lines (two of the six lines overlap with the materials for Theme 2). F_5 populations from the six groups are expected to be completed by 2025. However, only two ajara lines were used as founders in the current NAM population. To facilitate the development of highly efficient quinoa breeding utilizing ajara in the future, it will be necessary to establish a large-scale NAM population with ajara as the primary founder.

7.3 Developing Breeding Materials to Increase Quinoa's Resilience

In this section, we carried out the following studies on the development of early-maturing lines (Sect. 7.3.1) and the identification and functional analysis of genes that contribute to quinoa's resilience (Sect. 7.3.2).

7.3.1 The Development of Early-Maturing Lines

In the Altiplano, drought during the sowing season is becoming more serious, and what is most needed locally is an "early-maturing" quinoa variety that can produce a constant yield even if the sowing season is delayed due to drought. In this activity, "early maturity" has been identified as the top-priority trait for strengthening resilience, and Japanese and Bolivian research teams are working in parallel to develop markers and a breeding platform for the development of early-maturing quinoa varieties.

In Japan, based on the comprehensive integrated phenotypic and genotypic analysis data (Mizuno et al. 2020) of 136 quinoa inbred lines produced at JIRCAS, a Bolivian local line, which has large, white grains (Quinoa Real), was selected as a founder, and two early-maturing lines that contribute to strengthening resilience were selected as crossing breeding parents. Using the selected lines, two RIL populations were produced. Detailed phenotypic analyses were obtained for these F_6 generations (a total of approximately 350 lines) in a greenhouse. Of these, genotyping of F_6 generation plants was performed using dpMIG-seq for one of the crossing populations (approximately 160 lines), and phenotypic data was also obtained for the progeny F_7 plants. Based on these data, a big QTL involved in flowering was identified. Analysis using heterozygous residual lines around the QTL narrowed down the candidate QTL regions and identified markers involved in flowering. Further fine mapping and gene expression analysis are currently being used to identify the candidate gene. Furthermore, from the F_7 generation of the RIL population, we selected useful candidate breeding lines that have the early maturity QTL and exhibit large grain and short-stem traits.

In Bolivia, a local elite variety developed by PROINPA, which has large, white grains, was used as the founder. Early-maturing lines were selected as crossing breeding parents, including wild species, and RILs were produced at the Viacha PROINPA station (collaboration with activity 1). The three RILs population of F_4 or F_5 were grown in the Viacha field and evaluated the phenotype including flowering time maturing time, plant height, and seed size (Fig. 7.4). All of the lines were genotyped using dpMIG-seq. Based on QTL analysis of a RIL population with a wild species as one of the crossing parents, QTLs involved in early maturity were identified.

We are also developing useful breeding material that is adapted to the local environment. In addition to the problems of drought and salinity, there are also problems with damage to quinoa caused by mildew, frost, and hail. In order to produce quinoa lines with enhanced resilience to these biotic and abiotic stresses, we are also working on evaluating the traits and genotyping various breeding materials with different levels of resistance that have been selected in the field.

Fig. 7.4 Quinoa cultivation at the PROINPA experimental field in Viacha, Bolivia. Aerial view of the field in 2023. To avoid drought, an irrigation system has been installed in the experimental area of our project

7.3.2 Identification and Functional Analysis of Genes that Contribute to Quinoa's Resilience

Since quinoa is a plant with both male and female flowers on the same plant and is prone to hybridization, it is essential to create inbred lines with a fixed genetic background in order to conduct research in molecular biology and molecular genetics. Our research group has focused on this from an early stage, and we have disseminated our research to the world (Yasui et al. 2016). To date, we have created the largest number of inbred lines in the world, about 200 (submission in preparation), and analyzed the genotypic and phenotypic diversity of 136 of these lines, and the results show that the existing quinoa lines can be broadly classified into three groups: northern highland, southern highland, and lowland (Mizuno et al. 2020). We and the other research group have succeeded in decoding the whole genome of representative quinoa lines in these three groups at the chromosome level with high accuracy (Rey et al. 2023; Kobayashi et al. 2024a, b). We also developed a method for analyzing the functions of quinoa's endogenous genes by suppressing gene expression in quinoa itself or overexpression of foreign genes using a viral apple latent spherical virus (ALSV) (Ogata et al. 2021). Using the viral vector method, we have identified genes that control plant height and pigmentation from genomic information (Ogata et al. 2021). One of the most prominent morphological features of quinoa is that the aboveground portion is covered with a layer of epidermal bladder cells (EBCs). Recently we have found that EBCs play a role in water retention in quinoa leaves by suppressing two genes involved in EBC formation using a viral vector method (Fig. 7.5) (Kobayashi and Fujita 2024). We have now identified various candidate genes (such as flowering control genes and ion channel genes) that contribute to improving quinoa's resilience.

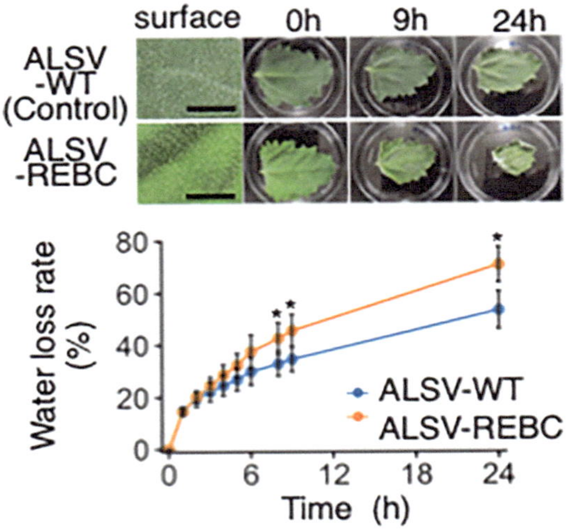

Fig. 7.5 Suppression of *Reduced number of EBC (REBC)* by the viral vector method significantly increases water loss in detached quinoa leaves. The photographs show changes over time in leaves detached from independent quinoa plants (inbred Iw line) inoculated with ALSV-REBC and ALSV-WT and dried on Petri dishes for 0, 9, and 24 h. Black bars indicate 5 mm. The graph shows the water loss rate of leaves detached from independent quinoa plants (inbred Iw line) inoculated with ALSV-REBC and ALSV-WT. The data show means±SD of water loss rates expressed as a percentage of the initial weights ($n = 4$ biologically independent replicates). (This figure is adapted from the reference Kobayashi and Fujita (2024))

Quinoa is more salt-tolerant than other crops, especially the southern highland type. To elucidate the mechanism, we are analyzing the salt stress response of quinoa. We have established an experimental system using 600 mM NaCl, which is equivalent to seawater, with seedlings and analyzed the accumulation of salt in the aboveground and underground areas. The results showed that southern highland quinoa lines grown around the Salar de Uyuni do not accumulate much salt in the aboveground, whereas lowland quinoa lines grown mainly in the coastal areas of Chile accumulate much salt in the aboveground area (Kobayashi et al. 2024b). We are currently identifying key genes for salt stress tolerance based on the results of comprehensive gene expression transcriptome analyses, as well as using QTL-seq methods to identify genes in the southern highland and lowland lines, which differ in the amount of salt accumulated in the aboveground area. In addition, to clarify the real situation of stress tolerance in quinoa in the fields around the Salar de Uyuni and, based on this situation, to establish the conditions for stress tolerance tests at the laboratory level, we are conducting a field sur-

Fig. 7.6 Quinoa grown in marginal environments in fields adjacent to the area where salt is accumulating at the edge of the Salar de Uyuni

vey in quinoa fields in the border regions adjacent to the Salar de Uyuni (Fig. 7.6).

Overall, the basis for molecular analysis has been established, and gene identification is now underway using a combination of forward and reverse genetics methods based on genome information. Based on the results, functional analysis of many genes has already been conducted. In addition, the project is leading the world in the identification and functional analysis of useful genes in quinoa and is attracting attention as a model case not only from quinoa researchers but also from researchers of other non-model plants, non-main crops, and orphan crops, and there are signs of subsequent developments.

7.4 The Development of Conservation and Management Techniques for Sustainable Quinoa-Based Agricultural Ecosystems Using Native Genetic Resources

In this section, we conducted the following studies on the development of a cultivation system that achieves sustained high productivity (Sect. 7.4.1), exploration and utilization of useful biological resources that contribute to quinoa production (Sect. 7.4.2), and improving the integrated quinoa-lama system (Sect. 7.4.3).

7.4.1 The Development of a Cultivation System that Achieves Sustained High Productivity

To develop sustainable cultivation techniques for quinoa, we are attempting to introduce leguminous crops into quinoa cropping systems. We are evaluating the effects of intercropping quinoa with native vetch in the northern region and with native perennial lupin in the southern region (Fig. 7.7). Not only will the nitrogen fixation by leguminous crops improve soil fertility, but we also expect that, particularly in the southern region, the lupins will cover the land after the quinoa harvest, significantly reducing soil erosion. To reduce the damage caused by soil erosion, we are promoting the use of native plants such as *Baccharis tola* as hedges to surround the perimeter of the fields. We are also developing a method to propagate these useful native plants.

Fig. 7.7 Quinoa and lupin mixed crop test field near the Salar de Uyuni

7.4.2 Exploration and Utilization of Useful Biological Resources that Contribute to Quinoa Production

To encourage farmers to adopt the sustainable quinoa production techniques introduced in Sect. 7.4.1, we are also examining whether we can add value to the native plants that we are trying to introduce into the quinoa cultivation system. We are particularly focusing on the secondary metabolites of native plants. In addition, we are also conducting research into the development of biological fertilizers and biological pesticides as technologies that contribute to organic farming.

7.4.3 Improving the Integrated Quinoa-Llama System

With a view to integrating agriculture and animal husbandry, we have established an experimental protocol in Japan for analyzing the reduction of lignin content in quinoa residues by inoculating white basidiomycetes into the stems of quinoa residues. As a preliminary test for analysis using domestic llamas, we conducted various experiments on alpacas, which are closely related to llamas and are raised in Hokkaido, and established methods for analyzing their nutritional status and determining pregnancy by analyzing their feces. The technology transfer of these national results began in 2022 at the Viacha Experimental Farm of PROINPA and UMSA.

For the sustainable production of quinoa, crop-livestock integration is an essential technology. The only livestock that can be raised locally are llamas, and their compost has long been returned to quinoa fields. However, with the rapid increase in the area of quinoa cultivation in recent years, there is a shortage of llama compost. To improve this situation, we are promoting the effective use of quinoa residues (stems) that have been left unused until now. Quinoa stems are hard, and llamas cannot eat them as they are, but by processing quinoa stalks with white basidiomycetes, llamas will be able to use the quinoa stalks. If this technology can be developed, it is expected that the nutritional status of llamas will improve and the amount of compost that can be returned to quinoa fields will increase, which will also contribute to improving quinoa productivity.

7.5 Building Technical Information Extension Networks in the Bolivian Altiplano

So far, we have selected two districts, Umala and Chacala, as model districts for technology dissemination. In these areas, a baseline survey was first conducted with 60 households in each area. The survey involved visiting each household and asking direct questions about household member characteristics, quinoa cultivation, llama husbandry, smartphone use, etc. As a result, the literacy rate of the 60 respondents in the Chacala area (including Chita and Keasa) was 100%, and the average family size was 4.3 people. The average cultivated area was 20.6 hectares, of which 10.3 hectares were planted with quinoa. 80% of the farmers practiced 1 year of fallow. 86% of the respondents used a smartphone, and 94% of them used WhatsApp (a smartphone SNS application with the same functions as LINE used in Japan) as a means of communication.

PROINPA held a technical training session for farmers in the Chacala region in March 2022. The workshop included practical training on quinoa cultivation techniques and soil conservation and was attended by about 40 farmers. During the workshop, we asked the farmers how they would like to receive technical information from the SATREPS Bolivia project in the future and found that 76% of the farmers would like to receive information directly at the workshop. It was also found that 61% would like to receive information via WhatsApp. In December 2022, we conducted additional interviews with ten households and found that there was a mix of full-time and part-time farmers among the quinoa farmers and that some produced and sold processed quinoa products (such as cakes and bread, saponin-removed quinoa, and quinoa flour). Given this situation of a diverse mix of farmers, a quinoa WhatsApp group was created around PROINPA to disseminate a wide range of information. Starting with the announcement of a local training session scheduled for March 2023 (Fig. 7.8), we started disseminating information through social networks using WhatsApp.

In Chita, a model district in the Uyuni region, PROINPA and local residents have collaborated to build a simple greenhouse for raising seedlings for hedges to prevent erosion. During a business trip in December 2022, we participated in a community meeting in Chita, and it was decided that from now on, based on this greenhouse, we would provide information on technology dissemination based on the needs of local residents.

7.6 Concluding Remarks

During the project period, the project was forced to deal with various external force majeure events, such as the Corona disaster,

Fig. 7.8 Farmers receiving an explanation from a technician about the characteristics of quinoa lines at a local training session held in Chita, Bolivia, in 2023

the decline in the value of currencies such as the yen and the Boliviano, and the deterioration of security following national elections, each time with the support of Bolivian and Japanese researchers, JST, JICA, embassies, and the Bolivian government. With this support, the project has managed to survive. On the other hand, droughts and other extreme weather events in Bolivia have continued year after year during the project period, further deepening our awareness that the results of our project are needed without waiting. The goal of the project is to contribute to food security and national income in Bolivia by developing and disseminating production technologies to increase the resilience of quinoa, a crop with high nutritional value. However, we have also learned that this project itself needs to strengthen its resilience. While strengthening the resilience of the project, we need to continue the project and achieve the necessary results.

Acknowledgments We especially thank G. Almanza, R. Oros, I. Morales, A. Bonifacio, W. Rojas, P. Mollinedo, J. Quezada, Y. Flores, J. Ciancas, G. Plata, S. Tarqui, and M. Villca for taking on important roles, including that of project leader. We also thank all the members of "SATREPS Bolivia Superfood Project," Universidad Mayor de San Andrés (UMSA), Fundación PROINPA (PROINPA), Kyoto University, Obihiro University of Agriculture and Veterinary Medicine, Tokyo University of Agriculture and Technology, and the Japan International Research Center for Agricultural Sciences (JIRCAS). This work was supported by Grants-in-Aid for Scientific Research (KAKENHI) from the Japan Society for the Promotion of Science (JSPS) (Grant Nos. JP21H02158 to YN, YF; JP16K07412, JP24510312 to YF); the Moonshot Research and Development Program for Agriculture, Forestry and Fisheries by Cabinet Office, Government of Japan [funding agency: the Bio-oriented Technology Research Advancement Institution (BRAIN); Grant No. JPJ009237]; the Science and Technology Research Partnership for Sustainable Development (SATREPS) of the Japan Science and Technology Agency (JST) and the Japan International Cooperation Agency (JICA) (Grant No. JPMJSA1907); and the Ministry of Agriculture, Forestry and Fisheries (MAFF) of Japan.

References

Kobayashi Y, Fujita Y (2024) Epidermal bladder cells play a role in water retention in quinoa leaves. Plant Biotechnol 41(4):447–452

Kobayashi Y, Hirakawa H, Shirakawa K et al (2024a) Chromosome-level genome assemblies for two quinoa inbred lines from northern and southern highlands of Altiplano where quinoa originated. Front Plant Sci 15:1434388

Kobayashi Y, Ryohei S, Fujita M et al (2024b) CqHKT1 and CqSOS1 mediate genotype-dependent Na^+ exclusion under high salt stress in quinoa. bioRxiv. https://doi.org/10.1101/2024.08.05.606677

Mizuno N, Toyoshima M, Fujita M et al (2020) The genotype-dependent phenotypic landscape of quinoa in salt tolerance and key growth traits. DNA Res 27:dsaa022

Mujica A, Jacobsen SE (2006) La quinua (*Chenopodium quinoa* Willd.) y sus parientes silvestres. In: Mónica Moraes R, Øllgaard B, Kvist LP, Borchsenius F, Balslev H (eds) Botánica Económica de los Andes Centrales, pp 449–457

Nishimura K et al (2024) Degenerate oligonucleotide primer MIG-seq: an effective PCR-based method for high-throughput genotyping. Plant J 118:2296–2317

Ogata T, Toyoshima M, Yamamizo-Oda C et al (2021) Virus-mediated transient expression techniques enable functional genomics studies and modulations of betalain biosynthesis and plant height in Quinoa. Front Plant Sci 12:643499

Rey E, Maughan PJ, Maumus F et al (2023) A chromosome-scale assembly of the quinoa genome provides insights into the structure and dynamics of its subgenomes. Commun Biol 6:1263

Wilson HD (1990) Quinua and Relatives (Chenopodium sect. Chenopodium subsect. Cellulata) Econo. Bot 44:92–110

Yasui Y, Hirakawa H, Oikawa T et al (2016) Draft genome sequence of an inbred line of Chenopodium quinoa, an allotetraploid crop with great environmental adaptability and outstanding nutritional properties. DNA Res 23:535–546

Yu J et al (2008) Genetic design and statistical power of Nested Association Mapping in Maize. Genetics 178:539–551

Development of New Vegetable Varieties for Sustainable Production in Tropical Regions: Tomato and Chili Pepper

8

Seung Won Kang, Ken Hoshikawa, Shigekazu Kawashima, Hiroshi Ezura, Naoya Fukuda, Koichi Sugimoto, Oscar Witere Mitalo, Nono Carsono, Farida Damayanti, Citra Bakti, Eliana Wulandari, Fitri Widiantini, and Zumi Saidah

Abstract

It is one of the urgent priorities to establish advanced breeding platforms for sustainable vegetable production in tropical regions. The overall goal of this project is to develop disease-resistant chili peppers and heat-tolerant tomatoes that can potentially improve productivity in Indonesia, one of the countries facing climate change in the tropical region. In addition, we will analyze the potential impact of new varieties for farmers, markets, and seed providers through socio-economic approaches. To accomplish these goals, our project is composed of four main research activities: (1) development of heat tolerant tomatoes through developing near isogenic lines (NILs), TILLING by sequencing platform and introducing genome editing technique, (2) development of anthracnose resistant chili pepper through collecting chili pepper germplasm in Indonesia, developing TILLING by sequencing platform using local varieties and evaluation and diagnostic techniques, dissemination of anthracnose resistant chili pepper developed by UNPAD, etc., (3) economic analysis and impact assessment evaluation of newly

S. W. Kang (✉) · H. Ezura · N. Fukuda · K. Sugimoto · O. W. Mitalo
Institute of Life and Environmental Sciences, Tsukuba Plant Innovation Research Center, University of Tsukuba, Tsukuba, Ibaraki, Japan
e-mail: kang.seungwon.ga@u.tsukuba.ac.jp; ezura.hiroshi.fa@u.tsukuba.ac.jp; fukuda.naoya.ka@u.tsukuba.ac.jp; sugimoto.koichi.gu@u.tsukuba.ac.jp; mitalo.witere.ke@u.tsukuba.ac.jp

K. Hoshikawa
Biological Resources and Post-harvest Division, Japan International Research Center for Agricultural Sciences, Tsukuba, Ibaraki, Japan
e-mail: hoshikawak0339@jircas.go.jp

S. Kawashima
Department of Integrative Studies of Plant and Animal Production, Miyagi University, Sendai, Miyagi, Japan
e-mail: kawashim@myu.ac.jp

N. Carsono · F. Damayanti · C. Bakti · E. Wulandari · F. Widiantini · Z. Saidah
Faculty of Agriculture, Universitas Padjadjaran, Jatinangor, Indonesia
e-mail: n.carsono@unpad.ac.id; farida.damayanti@unpad.ac.id; citra.bakti@unpad.ac.id; eliana.wulandari@unpad.ac.id; fitri.widiantini@unpad.ac.id; zumi.saidah@unpad.ac.id

T. Nagamine et al. (eds.), *Plant Genetic Resources (PGR) for Sustainable Crop Production*, Sustainable Development Goals Series, https://doi.org/10.1007/978-981-96-7117-5_8

developed varieties through socio-economic analysis, and (4) establishment of the Center of Advanced Breeding Technique (CAB-Tech) to facilitate advanced breeding of tomatoes and chili peppers as well as other important vegetables such as strawberries, develop supporting and training program for local farmers, researchers, students, etc.

Keywords

Anthracnose resistance · *Capsicum* spp. · *Colletotrichum* spp. · Ethyl methanesulfonate (EMS) · Genome editing · Heat tolerance · Mutation library · Near isogenic lines (NILs) · *Solanum lycopersicum* · TILLING by sequencing

8.1 Introduction

8.1.1 Background of the Project "Breeding Innovation in Chili Pepper and Tomato to Accelerate Sustainable Vegetable Production in Tropical Regions"

Indonesia, located in the Southeastern Asia of tropical region, is one of the largest countries not only with population of about 280 million but also 1,916,820 km^2 of its territory. In addition, Indonesia is one of the promising emerging markets and developing economies with stable GDP (1319.1billion US$ in 2024) growth at a rate of 5% every year. Agricultural industry is a key sector for economic development, meanwhile its contribution to GDP has declined from 40% in 1970s to 12.5% in 2023. Agriculture in Indonesia comprises smallholder production whose main products are rice, soybeans, corn, fruits, and vegetables including chili pepper and tomato, and large plantations producing palm oil and rubber.

Tomato and chili pepper are the most important fruit vegetables in Indonesia and considered the most desirable tropical vegetables for both the household and the industrial sector. In addition, their demand has been rapidly increasing accompanied by its population growth and huge economic growth. The production of tomato was ranked top 6th vegetable with 1.1 million tons accounting for 7.8% of vegetable production, while the main chili pepper species produced are *Capsicum annuum* and *C. frutescens* which account for 10.6 (1.55 million tons) and 10.3% (1.51 million tons), respectively. Therefore, the production of whole chili peppers occupies 21% of vegetable production with 3 million tons in Indonesia resulting in the top 3rd country of the pepper production in the world. The economic importance of chili pepper and tomato in Indonesia is so indispensable that, to support sustainable production of horticultural crops, Indonesian government has designated chili pepper as one of the top priorities in terms of food security and tomato as an additional important crop with considerable amount of production and consumption. Despite such efforts and supports by the government, both vegetables are facing various challenges such as low quality of seeds, low adoption of high yielding varieties, narrow genetic resources, low rate of releasing new superior varieties, short shelf life of harvested fruits, and pests and disease susceptibilities.

Climate change is one of the important factors raising the average global temperature by about 1.0 °C since 1880, and it is considered the global warming has started since 1970s, at a rate of roughly 0.11–0.22 °C per decade (Hansen et al. 2006). The impact of climate change on crop yields will be more severe in tropical regions (Cammarano et al. 2022). For example, at the end of 2016 to early 2017, the La Niña phenomenon impacted Indonesia with the intensity of rainfall higher than usual years resulting in the large price jump of chili pepper with more than three times caused by the high incidence of anthracnose disease on chili pepper, in addition to insect attacks and drought stress.

Development of new varieties is carried out by conventional breeding technique with a long period from 7 to 12 years or even more making it a time-consuming process with gaining low genetic variations. Recent advances in genomics, bioinformatics, transgenic, and genome editing have made it possible to accelerate the plant breeding process (Kumar et al. 2024; Li et al. 2022). The utilization of advanced techniques is essential to solve challenging problems in conventional breeding and to develop new varieties of heat tolerant tomato and disease resistant chili pepper. Therefore, advanced-breeding techniques such as molecular-genomic selection and genome editing (CRISPR/Cas9) are required not only to accelerate the breeding new cultivars but also to support sustainable production of tomato and chili pepper in Indonesia (Ramirez-Torres et al. 2021).

8.1.2 Objectives of the Project

It is one of the urgent priorities to establish advanced breeding platform for sustainable vegetable production in tropical regions not only in response to the rapid climate change and increasing demands of vegetable consumption but also in preparation for improving farmers' income and promoting public health.

The overall goal of this project "Breeding Innovation in Chili Pepper and Tomato to Accelerate Sustainable Vegetable Production in Tropical Regions," a five-year project which has been started from 2023 in Japan and Indonesia, is to develop disease resistant chili peppers and heat tolerant tomatoes that can potentially improve the productivity. In addition, we will analyze the potential impact of new varieties for farmers, markets, and seed providers through socio-economic approaches. Furthermore, to facilitate and implement our outputs to the Indonesian vegetable market, the Center of Advanced Breeding Technique (CAB-Tech) will be established at the Universitas Padjadjaran (UNPAD) in Indonesia.

8.2 Research Activities and Achievement of the Project

Our project is composed of four main research activities with several subactivities to accomplish the goals in the project as follows.

1. *Development of heat tolerant tomatoes* through developing near isogenic lines (NILs), Targeting Induced Local Lesions In Genomics (TILLING) by sequencing platform, genome editing, etc.
2. *Development of anthracnose resistant chili pepper* through collecting chili pepper germplasm in Indonesia, developing TILLING by sequencing platform using local variety and evaluation and diagnostic techniques, dissemination of anthracnose resistant chili pepper developed by UNPAD, etc.
3. *Economic analysis and impact assessment evaluation* of newly developed varieties by comparing farmers' preference and productivity with current varieties used.
4. *Establishment of Center of Advanced Breeding Technique (CAB-Tech)* to facilitate advanced breeding of tomatoes and chili peppers as well as other important vegetables such as strawberries, develop supporting and training program for local farmers, researchers, students, etc.

8.2.1 Development of Heat Tolerant Tomatoes in Tropical Regions

Tomato Production in Indonesia

Tomatoes are one of the important fruit vegetables in Indonesia, significantly contributing to food security and the national economy (Surni et al. 2020). However, tomato cultivation in Indonesia is increasingly threatened by climate change and global warming, exacerbating existing challenges and introducing new uncertainties for farmers (Kuswardhani et al. 2013). Tomatoes are widely cultivated across Indonesia, particularly in Java, Sumatra, and Sulawesi regions, with the highest tomato production in 2023 from West Java province (Statistics Indonesia 2024). Tomato production also serves as an important source of livelihood for smallholder farmers. These farmers typically have small plots of land, which they can utilize intensively to grow tomatoes using existing labor. Since tomatoes are a short-duration crop, they provide a consistent income that meets the daily necessities of farming households. Smallholder farmers are often poor and food-insecure, making income from tomato production vital for improving food and nutrition security and alleviating poverty (Zani et al. 2019).

The most common tomato varieties grown include "Roma," "Cherry," and other hybrid tomatoes, chosen for their adaptability to diverse microclimates. Despite the availability of high-yield seeds, farmers often face productivity losses due to pests, diseases, and the adverse effects of climate change. Indonesia is among the nation's most vulnerable to climate change. Rising temperatures, erratic rainfall, and frequent extreme weather events directly impact tomato production (Hoshikawa et al. 2021). Key challenges for production of tomatoes include rising temperatures (Hoshikawa et al. 2021), erratic rainfall patterns (Mahrup et al. 2023), increased pest and disease pressure (Tsai et al. 2006), soil degradation and erosion (Rudiarto and Doppler 2013), and functional shift of farmland (Rahman et al. 2024).

Various efforts are underway to mitigate the impact of climate change on tomato farming in Indonesia, including research and development (Dhugga 2022; Mubarok et al. 2023; Sakya et al. 2020; Szurman-Zubrzycka et al. 2023), climate-smart agricultural practices (Retnaningtyas et al. 2024), policy support and funding (Yijo et al. 2021), and global and local collaborations (Kuswardhani et al. 2013).

Current Outputs in Development of Heat Tolerant Tomato

Efforts to overcome the challenges of tomato production in Indonesia are being advanced through the Science and Technology Research Partnership for Sustainable Development (SATREPS) project, a collaborative initiative between UNPAD in Indonesia and the University of Tsukuba (UT) in Japan. This project employs several strategies to develop heat-tolerant tomato varieties by developing NILs by hybridizing Indonesian tomato varieties with heat-tolerant mutants lines obtained by the TILLING platform based on Micro-Tom at the University of Tsukuba, establishing a TILLING by sequencing platform, applying gene editing technologies, and optimizing cultivation technologies. Mutant libraries of tomatoes are one of the essential sources for breeding in addition to the natural variations among varieties or wild relatives. The mutants are also useful for gene discovery and functional analyses of the genes. The University of Tsukuba has developed the tomato mutant collections induced by ethyl methanesulfonate (EMS) treatment and γ-ray irradiation, as a part of National BioResource Project supported by the Ministry of Education, Culture, Sports, Science and Technology (MEXT), Japan. The collected mutants are open and available via TOMATOMA database (Saito et al. 2011; Shikata et al. 2016; https://tomatoma.nbrp.jp/). One accession of the mutants harbored, on average, more than 600 mutations, such as single nucleotide variants, insertions, and deletions (Shirasawa et al. 2016). Specific mutations in the gene of interest are detectable by TILLING screening with the bulked genomic DNA of mutants for reverse genetics (Okabe et al. 2011). More recently, large scale 2nd/3rd generation sequencing provides genome-wide mutation dataset for in silico TILLING platform (Gupta et al. 2017; Yano et al. 2019). Various types of tomato mutants involving in parthenocarpic fruit set (Okabe et al. 2019; Schubert et al. 2019), fruit maturation (Mubarok et al. 2019), and growth habit suitable for indoor cultivation system (Kwon et al. 2020) have been reported using these platforms.

To develop NILs, we evaluated fruit quality and yields in six Indonesian tomato varieties for the recurrent parental lines in NILs development and three out of six varieties were selected based on parameters such as fruit size, fruit weight, carotenoid, and so on (Data are not shown). We hybridized a heat tolerant mutant line of tomato Micro-Tom with above three varieties to proceed generations. In addition, we evaluated phenotypic performance of fourteen commercial varieties under heat stress conditions (Fig. 8.1). Furthermore, we could successfully amplify 38 out of 45 simple sequence repeat (SSR) markers evaluated from a total of eight different tomatoes: six commercial lines used to develop NILs, one heat tolerant mutant line, and a wild type of Micro-Tom. We used EMS, a chemical mutagen-

Fig. 8.1 Germplasm of tomatoes used to evaluate phenotypic performance under heat stress conditions

esis, to develop a mutant library of Micro-Tom combined with TILLING by sequencing platform. We have obtained more than 2,000 of M_1 generation and are working on generating mutants to obtain enough number of mutants to potentially cover whole genome range of Micro-Tom (Fig. 8.2).

8.2.2 Development of Anthracnose-Resistant Chili Pepper in Tropical Regions

An Overview of Chili Pepper in Indonesia
Chili (*Capsicum* spp.) is one of the most important horticultural crops designated by the Indonesian government as an essential necessity item based on Presidential Decree No. 59 of 2020 and cultivated widely, as it is not only essential in local cuisine but also important contributor for the agricultural economy. Therefore, this commodity has a high household expenditure allocation and affects the inflation rate (0.15% in 2019). About 67% of chilies produced is consumed fresh by the Indonesian people as a spice and food flavoring, meanwhile only 14% is used for industrial purposes.

The increasing demand for chili peppers, both for domestic consumption (2.0 kg per capita and 2.2 kg of cayenne pepper; Centre for Agricultural Data and Information System 2023a) and export (increase by 4.9% in 2022; Centre for Agricultural Data and Information System 2023b), has led to intensified agricultural practices. However, these practices have also heightened the vulnerability of chili crops to various diseases. Four kinds of diseases are the most dominant causes of low chili production in Indonesia, i.e., anthracnose (*Colletotrichum* spp.), blight (*Phytophthora capsici*), bacterial wilt (*Ralstonia solanacearum*), and viruses such as Cucumber mosaic virus (CMV), Potato virus (Y/PVY), etc. (Suryaningsih et al. 1996).

Anthracnose disease is one of the major constraints in many chili producing countries including in Indonesia. Chili peppers can be infected in all growing stages showing dark, sunken lesions on chili fruits, causing the chili fruits to be unmarketable and resulting in substantial economic losses (Mongkolporn et al. 2010; Saxena et al. 2016; Than et al. 2008a). The productivity of chili peppers in Indonesia is still low compared to the yield potential, ranging from 2.96 to 7.39 tons·ha^{-1}, with an average productivity of 6.84 tons·ha^{-1} (Syukur et al. 2010). Yield loss in the field due to anthracnose disease in the rainy season reaches 75–80%, while in the dry season reaches 20–35% (Kusandriani and

Fig. 8.2 Expansion of Micro-Tom M_1 generation in the net house at BPSITS Lembang, Indonesia

Permadi, 1996). A total of 24 species of *Colletotrichum* have been identified as the causal agent of chili anthracnose, with three main species found in Indonesia: *C. acutatum*, *C. truncatum*, and *C. gloeosporioides* (Mongkolporn and Taylor 2018; Voorrips et al. 2004).

The use of chili varieties with genetic resistance to anthracnose represents a promising strategy for sustainable disease management. Utilization of resistant varieties is the most cost-effective strategy to control anthracnose disease in chili peppers (Kiran et al. 2020). Due to the lack of resistance in the *C. annuum* gene pool, an introgression gene from another species is needed (Temiyakul et al. 2012). Some chili pepper genotypes resistant to *Colletotrichum* spp. have been reported, and four genotypes such as Lembang 1, Tanjung 2, Hiyung, and Tanjung in Indonesia have been known to have moderate anthracnose resistance (Adhni et al. 2022). Tanjung 2 is one of the national varieties cultivated in Indonesia. It has a compact plant architecture, medium tall, beautifully red fruit, can be harvested 58 days after planting, is rather tolerant to anthracnose, and has a yield potential of 12 tons·ha^{-1}.

Developing molecular markers linked to resistance genes are one of the widely used molecular breeding techniques, allowing for selecting resistant varieties at an earlier stage in the breeding process. Nugroho et al. (2019) have identified 14 molecular markers which enable to distinguish 15 chili pepper varieties of *C. annuum* and *C. frutescens*. Only one marker, GPMS29, showed a high significant association with anthracnose resistance.

Challenge to Develop Chili Pepper Resistant to Anthracnose

The development of new chili pepper varieties with high yielding potential, resistant to pathogens and diseases, and adapted to climate change through advanced plant breeding programs are urgently required. To facilitate anthracnose resistant chili pepper, we have set up four main activities: (a) exploration and collection of local chili pepper germplasms, (b) dissemination of UNPAD's chili pepper cultivars, (c) development of mutant chili pepper populations using the TILLING platform, and (d) development of selection techniques by phenotyping and genotyping of the resistance genes to anthracnose. Main target genes to develop anthracnose resistant chili pepper will be *RCt1*, *co1*, *co2*, *co3*, *co4*, *Co5*, or other candidate genes related to the capsaicin biosynthesis and cuticle thickness. The *RCt1* gene was detected on *C. annuum* cv. Punjab Lal from India, which is resistant to *C. truncatum* showing correlation between capsaicin content and cuticle thickness and the resistance to anthracnose. The other *co1*, *co2*, *co3*, *co4*, or *Co5* genes were identified from *C. baccatum* "PBC80" and *C. chinense* "PBC932" showing resistance to *C. capsici* and *C. acutatum* (De Silva et al. 2017; Mongkolporn et al. 2010; Ranathunge et al. 2012; Than et al. 2008b).

Current Research Outputs

To develop anthracnose-resistant chili pepper lines, we have been collecting chili pepper germplasm from eight regions in Indonesia (Fig. 8.3). We have collected local and commercial varieties (Data are not shown) (Fig. 8.4). Second, we have introduced chemical mutagenesis to create mutation population by treating EMS to more than 6,000 seeds of Tanjung 2 and now we harvested and expanded M_2 generation seeds. A TILLING by sequencing technique will be used to detect mutations from M_2 chili pepper generation of Tanjung 2. Finally, we have been collected more than one hundred isolates of *Colletotrichum* spp. from six provinces of major chili producers, and the top five most virulent isolates from each region were assessed for their sensitivity against commonly used fungicides (data are not shown).

8.2.3 Evaluation on Economic Analysis and Impact Assessment of Newly Developed Varieties of Tomatoes and Chili Pepper

The applicability of excellent lines of tomato and chili pepper is evaluated through socio-economic analysis. The socio-economic study is performed at Universitas Padjadjaran in collaboration with experts from the UT and Miyagi University. The

Fig. 8.3 Target regions to collect domestic germplasms of chili peppers in Indonesia

Fig. 8.4 Example of phenotypic traits of collected germplasms of chili pepper

team of Indonesia and Japan have been working collaborative research to analyze the main data.

The activities during the project especially for Output 3 are as follows:

1. Conducting economic survey of farmers and on quality assessment of seeds used by farmers
2. Performing market survey on tomato and chili pepper
3. Collecting farm management data (yield, income, costs, etc.) of cultivation techniques of promising lines from small and medium scale farmers
4. Performing consumers' preference for excellent lines of tomatoes and chili peppers

Farmer economic survey focuses on identifying seed sources, varieties, seed prices, yields, sales prices, and sales destinations and to evaluate the quality of seeds used by farmers indicated by germination rates. Four districts in West Java were selected as the study area considering as the centers of chili and tomato production, i.e., Garut, Bandung, Cianjur, and Ciamis Districts. In general, the farmers grew three types of chilies, i.e., big chili (*Capsicum annuum*), curly chili (*C. annuum*), and cayenne pepper (*C. frutescens*). Big chili is 1.8–2.0 cm in diameter, around 18 cm of length, and 19–20 g of fresh weight. Curly chili is 0.6–0.8 cm in diameter, 17–19 cm of length, and 6–7 g of fresh weight. Cayenne pepper is around 0.5 cm in diameter, 4.5 cm of length, and 2.5 g of fresh weight.

For tomatoes, the farmers grow two types of tomatoes, i.e., big tomato called as "tomat buah" (*Solanum lycopersicum*) and vegetable or small tomato called as "tomat sayur" (*S. lycopersicum*). Vegetable tomato is smaller and round with a diameter of 2–4 cm while big tomato is bigger, oval, and 5–10 cm in diameter. Apart from that, vegetable tomatoes usually have a sharper and sour taste, while big tomatoes tend to be sweeter.

The survey of farmers and traders were conducted to identify to which markets the selling of tomato and chili pepper, for instance, those products were sold to several markets such as to middleman, traditional market, and horeka (hotel, restaurant, and café). Each market has its preference due to the quality of tomato and chili pepper depending on their types. For instance, the preference of curly chili pepper would be fresh, straight, shiny, and red color. For small chili, the markets prefer fresh, red color straight, and thick, while for big chili, the markets prefer fresh, green color, straight and big. For tomatoes, the preference of big tomato would be fresh, oval, big, and shiny red color, while for small tomato (tomat sayur), the market prefers fresh, round, and shiny red color.

The collection of the farm management data focuses on yield, income, costs, and cultivation techniques. The farm management data refers to small and medium scale farmers since most of the farmers in Indonesia come under this categorization. The measurement is performed by calculating cost efficiency, such as cost efficiency analysis, productivity, and profitability analysis.

The activity of consumers' preference includes the identification of price, size, taste such as acidity and sweetness, texture, and so on. The consumers' preference data is analyzed using statistical analysis. Performance analysis, which represents the impact assessment of this project, is also assessed to measure the satisfaction of products, facilities, services, and assistance in using the products. The information presented is in the form of an ordinal measurement scale using a Likert scale, which has a scale range from 1 to 5. A score of 5 is strongly agree/very important and 1 is strongly disagree/very unimportant. The Importance Performance Analysis (IPA) is a technique aimed at the relationship between respondents' perceptions and priorities for the improvement of product; this relation is known as quadrant analysis (Martilla and James 1977). This analysis has been used in various fields of study because of its ease of implementation and the display of the results that facilitate proposed performance improvements (Martinez 2003). IPA analysis has the main function to display information related to service factors that affect respondents' satisfaction and loyalty, as well as priority service factors that need to be improved.

8.2.4 Establishment of the Center of Advanced Breeding Technique (CAB-Tech) for Sustainable Vegetable Production in Tropical Countries

The Objectives of the CAB-Tech in Relation to the Sustainable Vegetable Production

Since April 2024, the Center of Advanced Breeding Technique (CAB-Tech) has been established to facilitate the development of new vegetable varieties using an advanced technology and innovation in the field of plant breeding by aiming a core center of advanced breeding of vegetables and other crops in tropical region at the Faculty of Agriculture, Universitas Padjadjaran. We are expecting the plant breeding process of important vegetable crops such as tomatoes, chili peppers, and strawberries can be accelerated by sharing bioresources and introducing advanced breeding techniques through CAB-Tech not only in Indonesia but also in other countries in the tropical regions by increasing resource efficiency, the effectiveness of achieving plant breeding programs, increasing genetic progress (genetic gain), as well as shortening time required to create new of important crops.

The CAB-Tech is focusing on the development of new responding to climate changes to meet the needs of stakeholders, farmers, consumers, government, etc. In addition to this, we will develop training program not only for farmers or young frontiers who potentially contribute to improve productivity of horticultural crops in Indonesia but also for young scientists who will be in the fields of molecular breeding, bioinformatics, genomics, genetic engineering, and so on. The activity of CAB-Tech includes farmer communities' involvement through short- and long-term training programs in handling seeds of new varieties, genotypic evaluation towards the end-users' preferences (i.e., farmers and consumers), and informal seed multiplication.

We aim to (1) produce new that meet the needs of users such as farmers, seed companies, (2) shorten a plant breeding program utilizing advanced breeding techniques, (3) accelerate dissemination of new varieties of important horticultural crops and other crops, (4) implement advanced breeding technology for horticultural crops in tropical regions, and (5) strengthen link CAB-Tech activity to domestic and international research institutes, producers, and other stakeholders.

8.3 Challenges and Future Prospects

To produce and distribute enough resources such as food, feed, fiber, and fuel to meet the drastically increasing demands caused by growing population is the first and foremost continuing challenge faced by the agricultural sector in tropical countries including Indonesia. Indonesia has been experiencing significant economic growth at a rate of 5% of GDP increase every year. In recent, the middle class is expanding as their incomes rise, resulting in change of preference of food from rice, an essential staple for Indonesians, to noodles, bread, functional foods or health-related foods, etc. Average increase of vegetable production in Indonesia was 2.7% every year so production increased from 7.73 million tons to 13.8 million tons between 2003 and 2013 (AIP-PRISMA 2017). On the other hand, vegetable consumption per capita has declined as Indonesian people take only an average half amount of daily necessary vegetables.

In addition, the agricultural sector in Indonesia has been required not only to improve productivity but also to develop new varieties that can meet increasing Indonesian demands. Furthermore, the agricultural sector is also required to contribute to the eradication of poverty and hunger, achievement of food security, improvement of nutrition, promotion of sustainable agriculture and economic prosperity, and preservation of natural resources and human welfare.

Outputs expected through our project are heat-tolerant tomato developed by genome editing, promising lines of tomato and chili pepper, protocol of advanced tomato cultivation, chili germplasm collecting local varieties in Indonesia, mutant population of tomato and chili pepper by TILLING by sequencing platform, economic analysis, and impact assessment of new tomato and chili pepper varieties, etc.

We expect outputs generated through the SATREPS project can contribute to improve the productivity of important vegetables in terms of food security in response to climate change and the spread of plant disease in tropical regions by achieving three sustainable development goals (SDGs) "2. ZERO HUNGER," "13. CLIMATE CHANGE," and "15. LIFE ON LAND."

References

Adhni AL, Fitriyanti D, Liestiany E (2022) Uji Ketahanan beberapa varietas cabai (*Capsicum* sp.) terhadap Penyakit Antraknosa (*Colletotrichum* sp.) yang berasal dari Desa Hiyung Kabupaten Tapin. Proteksi Tanaman Tropika 5:448–454

AIP-PRISMA (2017, Sept) Vegetable Sub-Sector Growth Strategy Document for East Java (EJ)

Cammarano D, Jamshidi S, Hoogenboom G, Ruane AC, Niyogi D, Ronga D (2022) Processing tomato production is expected to decrease by 2050 due to the projected increase in temperature. Nat Food 3:437–444. https://doi.org/10.1038/s43016-022-00521-y

Centre for Agricultural Data and Information System (2023a) Statistics of food consumption. Secretariate General Ministry of Agriculture Indonesia

Centre for Agricultural Data and Information System (2023b) Market performance analysis of chili. Secretariate General Ministry of Agriculture Indonesia

De Silva DD, Ades PK, Crous PW, Taylor PWJ (2017) *Colletotrichum* species associated with chili anthracnose in Australia. Plant Pathol 66:254–267. https://doi.org/10.1111/ppa.12572

Dhugga KS (2022) Gene editing to accelerate crop breeding. Front Plant Sci 13:889995. https://doi.org/10.3389/fpls.2022.889995

Gupta P, Reddaiah B, Salava H, Upadhyaya P, Tyagi K, Sarma S, Datta S, Malhotra B, Thomas S, Sunkum A, Devulapalli S, Till BJ, Sreelakshmi Y, Sharma R (2017) Next-generation sequencing (NGS)-based identification of induced mutations in a doubly mutagenized tomato (*Solanum lycopersicum*) population. Plant J 92:495–508. https://doi.org/10.1111/tpj.13654

Hansen J, Sato M, Ruedy R, Lo K, Lea DW, Medina-Elizade M (2006) Global temperature change. Proc Natl Acad Sci 103:14288–14293. https://doi.org/10.1073/pnas.0606291103

Hoshikawa K, Pham D, Ezura H, Schafleitner R, Nakashima K (2021) Genetic and molecular mechanisms conferring heat stress tolerance in tomato plants. Front Plant Sci 12:786688. https://doi.org/10.3389/fpls.2021.786688

Kiran R, Akhtar J, Kumar P, Shekhar M (2020) Anthracnose of Chilli: status, diagnosis, and management. IntechOpen. https://doi.org/10.5772/intechopen.93614

Kumar R, Das SP, Choudhury BU, Choudhury BU, Kumar A, Prakash NR, Verma R, Chakraborti M, Devi AG, Bhattacharjee B, Das R, Das B, Devi HL, Das B, Rawat S, Mishra VK (2024) Advances in genomic tools for plant breeding: harnessing DNA molecular markers, genomic selection, and genome editing. Biol Res 57:80. https://doi.org/10.1186/s40659-024-00562-6

Kusandriani Y, Permadi H (1996) Pemuliaan tanaman cabai. In: Duriat AS, Widjaja A, Hadisoeganda W, Soetiarso TA, Prabaningrum L (eds) Teknologi Produksi Cabai Merah. Balai Penelitian Tanaman Sayuran, Lembang, pp 28–35

Kuswardhani N, Soni P, Shivakoti GP (2013) Comparative energy input-output and financial analyses of greenhouse and open field vegetable production in West Java, Indonesia. Energy 53:83–92. https://doi.org/10.1016/j.energy.2013.02.032

Kwon CT, Heo J, Lemmon JZH, Capua Y, Hutton SF, Van Eck J, Park SJ, Lippman ZB (2020) Rapid customization of Solanaceae fruit crops for urban agriculture. Nat Biotechnol 38:182–188. https://doi.org/10.1038/s41587-019-0361-2

Li W, Liu J, Zhang H, Liu Z, Wang Y, Xing L, He Q, Du H (2022) Plant pan-genomics: recent advances, new challenges, and roads ahead. J Gen Genom 49:833–846. https://doi.org/10.1016/j.jgg.2022.06.004

Mahrup M, Kusnartha IGM, Sukartono S, Silawibawa P, Susilowati LE, Padusung P, Fahrudin F (2023) Adaptive parameters for smart climate agriculture: a case study on erratic climate in Lombok, Eastern Indonesia. IOP Conf Ser Earth Environ Sci 1183:012023

Martilla JA, James JC (1977) Importance-performance analysis. J Market 41:77–79

Martinez CL (2003) Evaluation report: tools cluster networking meeting# 1. CenterPoint Institute, Inc, Arizona

Mongkolporn O, Montri P, Supakaew T, Taylor PWJ (2010) Differential reactions on mature green and ripe chili fruit infected by three *Colletotrichum* spp. Plant Dis 94:306–310

Mongkolporn O, Taylor PWJ (2018) Chili anthracnose: *Colletotrichum* taxonomy and pathogenicity. Plant Pathol 67:1255–1263. https://doi.org/10.1111/ppa.12850

Mubarok S, Hoshikawa K, Okabe Y, Yano R, Tri MD, Ariizumi T, Ezura H (2019) Evidence of the functional role of the ethylene receptor genes *SlETR4* and *SlETR5* in ethylene signal transduction in tomato. Mol Gen Genomics 294:301–313. https://doi.org/10.1007/s00438-018-1505-7

Mubarok S, Jadid N, Widiastuti A, Derajat Matra D, Budiarto R, Lestari FW, Nuraini A, Suminar E, Rahmat BPN, Ezura H (2023) Parthenocarpic tomato mutants, *iaa9-3* and *iaa9-5*, show plant adaptability and fruiting ability under heat-stress conditions. Front Plant Sci 14:1090774. https://doi.org/10.3389/fpls.2023.1090774

Nugroho K, Terryana RT, Manzila I, Priyatno TP, Lestari P (2019) The use of molecular markers to analyse the genetic diversity of Indonesian pepper (*Capsicum* spp.) varieties based on anthracnose resistance. Makara J Sci 23:137–147. https://doi.org/10.7454/mss.v23i3.11261

Okabe Y, Asamizu E, Saito T, Matsukura C, Ariizumi T, Brès C, Rothan C, Mizoguchi T, Ezura H (2011) Tomato TILLING technology: development of a reverse genetics tool for the efficient isolation of mutants from Micro-Tom mutant libraries. Plant Cell Physiol 52:1994–2005. https://doi.org/10.1093/pcp/pcr134

Okabe Y, Yamaoka T, Ariizumi T, Ushijima K, Kojima M, Takebayashi Y, Sakakibara H, Kusano M, Shinozaki Y, Pulungan SI, Kubo Y, Nakano R, Ezura H (2019) Aberrant stamen development is associated with parthenocarpic fruit set through up-regulation of gibberellin biosynthesis in tomato. Plant Cell Physiol 60:38–51. https://doi.org/10.1093/pcp/pcy184

Rahman M, Subiyakto B, Rusmaniah J, Ilhami MR (2024) Factors caused the convertion of agricultural and into a residential area in Belayung Baru village. Innov Soc Stud J 6:28–35

Ramirez-Torres F, Ghogare R, Stowe RE, Cerdá-Bennasser P, Lobato-Gómez M, Williamson-Benavides BA, Giron-Calva PS, Hewitt S, Christou P, Dhingra A (2021) Genome editing in fruit, ornamental, and industrial crops. Transgenic Res 30:499–528. https://doi.org/10.1007/s11248-021-00240-3

Ranathunge NP, Mongkolporn O, Ford R, Taylor PWJ (2012) *Colletotrichum truncatum* pathosystem on *Capsicum* spp: infection, colonization and defence mechanisms. Aus Plant Pathol 41:463–473. https://doi.org/10.1007/s13313-012-0156-0

Retnaningtyas TA, Padmaningrum D, Anantanyu S (2024) Implementation of Climate-Smart Agriculture (CSA) practices for sustainable agriculture development by millennial farmer in West Java Province (case study in Nudira Fresh, PT Nudira Sumber Daya Indonesia). In: Proceeding of the International conference on multidisciplinary studies (ICOMSI), vol 1, pp 7–13

Rudiarto I, Doppler W (2013) Impact of land use change in accelerating soil erosion in Indonesian upland area: a case of Dieng Plateau, Central Java-Indonesia. Int J Agric Sci 3:558–576

Saito T, Ariizumi T, Okabe Y, Asamizu E, Hiwasa-Tanase K, Fukuda N, Mizoguchi T, Yamazaki Y, Aoki K, Ezura H (2011) TOMATOMA: a novel tomato mutant database distributing Micro-Tom mutant collections. Plant Cell Physiol 52:283–296. https://doi.org/10.1093/pcp/pcr004

Sakya AT, Sulistyaningsih E, Purwanto BH, Indradewa D (2020) Drought tolerant indices of lowland tomato cultivars. Indo J Agri Sci 21:59–69

Saxena S, Raghuwanshi R, Gupta VK, Singh HB (2016) Chili anthracnose: the epidemiology and management. Front Microbiol 30:527. https://doi.org/10.3389/fmicb.2016.01527

Schubert R, Dobritzsch S, Gruber C, Hause G, Athmer B, Schreiber T, Marillonnet S, Okabe Y, Ezura H, Acosta IF, Tarkowska D, Hause B (2019) Tomato MYB21 acts in ovules to mediate jasmonate-regulated fertility. Plant Cell 31:1043–1062. https://doi.org/10.1105/tpc.18.00978

Shikata M, Hoshikawa K, Ariizumi T, Fukuda N, Yamazaki Y, Ezura H (2016) TOMATOMA update: phenotypic and metabolite information in the Micro-Tom mutant resource. Plant Cell Physiol 57:e11. https://doi.org/10.1093/pcp/pcv194

Shirasawa K, Hirakawa H, Nunome T, Tabata S, Isobe S (2016) Genome-wide survey of artificial mutations induced by ethyl methanesulfonate and gamma rays in tomato. Plant Biotechnol J 14:51–60. https://doi.org/10.1111/pbi.12348

Statistics Indonesia (2024) Production of vegetables, 2021–2023. https://www.bps.go.id/en/statistics-table/2/NjEjMg==/produksi-tanaman-sayuran.html. Retrieved on December 5, 2024

Surni S, Saediman H, Wulandari F, Zani M, Yunus L, Taridala SAA (2020) Profitability and constraints of small-scale tomato production in Baubau municipality of Southeast Sulawesi. WSEAS Trans Environ Dev 16:219–225. https://doi.org/10.37394/232015.2020.16.22

Suryaningsih ER, Sutarya, Duriat AS (1996) Penyakit tanaman cabai merah dan pengendaliannya. In: Duriat AS, Widjaya A, Thomas WH, Prabaningrum L (eds) TeknologiProduksi Cabai Merah. BalaiPenelitian Sayuran, Lembang (ID), pp 64–83

Syukur M, Sujiprihati S, Siregar A (2010) Pendugaan parameter genetic beberapa karakter agronomi cabai F4 dan evaluasi daya hasilnya menggunakan Rancangan Perbesaran (Augmented Design). J Agrotropika 1:9–16

Szurman-Zubrzycka M, Kurowska M, Till BJ, Szarejko I (2023) Is it the end of TILLING era in plant science? Front Plant Sci 14:1160695. https://doi.org/10.3389/fpls.2023.1160695

Temiyakul P, Taylor PW, Mongkolporn O (2012) Differential fruit maturity plays an important role in chili anthracnose infection. J King Mongkut's Univ Technol North Bangkok 22:494–504

Than PP, Prihastuti H, Phoulivong S, Taylor PWJ, Hyde KD (2008a) Chilli anthracnose disease caused by *Colletotrichum* species. J Zhejiang Univ Sci B 9:764–778. https://doi.org/10.1631/jzus.B0860007

Than PP, Rajesh J, Hyde KD, Pongsupasamit S, Mongkolporn O, Taylor PWJ (2008b) Characterization and pathogenicity of *Colletotrichum* species associated with anthracnose infection on chili (*Capsicum* spp.). Plant Pathol 57:562–572

Tsai WS, Shih SL, Green SK, Rauf A, Hidayat SH, Jan F-J (2006) Molecular characterization of Pepper yellow leaf curl Indonesia virus in leaf curl and yellowing diseased tomato and pepper in Indonesia. Plant Dis 90:247–247. https://doi.org/10.1094/pd-90-0247b

Voorrips RE, Finkers R, Sanjaya L, Groenwold R (2004) QTL mapping of anthracnose (*Colletotrichum* spp.) resistance in a cross between *Capsicum annuum* and *C. chinense*. Theo App Gen 109:1275–1282. https://doi.org/10.1007/s00122-004-1738-1

Yano R, Hoshikawa K, Okabe Y, Wang N, Dung PT, Imriani PS, Shiba H, Ariizumi T, Ezura H (2019) Multiplex exome sequencing reveals genome-wide frequency and distribution of mutations in the 'Micro-Tom' Targeting Induced Local Lesions in Genomes (TILLING) mutant library. Plant Biotechnol 36:223–231. https://doi.org/10.5511/plantbiotechnology.19.0830a

Yijo S, Asnawati A, Darma S, Achmad GN, Arizandi MA, Hidayati T, Darma DC (2021) Social experiments on problems from tomato farmers during Covid-19-Indonesia case. SAR J Sci Res 4:7–13. https://doi.org/10.18421/SAR41-02

Zani M, Saediman H, Abdullah S, Daud L, Yunus L (2019) Determinants of household food expenditure in a cassava-growing village in Southeast Sulawesi. Acad J Interdisc Stud 8:302–310. https://doi.org/10.36941/ajis-2019-0028

Research on the Practical Application of Grassland Plants for Restoring Degraded Grasslands and Promoting Healthy Growth of Livestock in Mongolia

9

Toshihiro Murata, Takeshi Nakano, Toshiya Okuro, Kenroh Sasaki, Masumi Ebina, and Tadao Asami

Abstract

Climate change has caused global food insecurity owing to extreme weather conditions, desertification, water shortages, and soil degradation. In Mongolia, global warming and the rapid increase in livestock numbers from 23 to 66 million over the past 28 years have led to severe overgrazing and degradation of Mongolian grasslands. This degradation worsens during harsh winters, causing frequent grass shortages that make it challenging for nomadic herders to sustain their livelihoods, and exacerbates poverty and other environmental issues. To address grassland degradation and maintain livestock health, efforts are being made to restore ecosystems and revitalize the livestock industry by utilizing native Mongolian plants through molecular biology, breeding, pharmacology, and sustainable agricultural methods. Although annual plants are effective for short-term recovery, maintaining the long-term health of grasslands and livestock requires the sowing of perennial plants, high-yield forage grasses, and the intercropping of high-performance plants. The ultimate goal is to distribute seeds and cultivation systems across grasslands, restore nomadic lifestyles, and contribute to Mongolia's societal development. The project comprises three interrelated themes, each led by experts, and focuses on selecting and analyzing forage grasses to sustainably improve plant productivity, exploring forage grasses that enhance livestock functions, support nomadic herders, and overcome grassland degradation.

T. Murata · K. Sasaki
Faculty of Pharmaceutical Sciences, Tohoku Medical and Pharmaceutical University, Sendai, Miyagi, Japan
e-mail: murata-t@tohoku-mpu.ac.jp; kenrs@tohoku-mpu.ac.jp

T. Nakano
Graduate School of Biostudies, Kyoto University, Kyoto, Japan
e-mail: Nakano.takeshi.6x@kyoto-u.ac.jp

T. Okuro · T. Asami (✉)
Graduate School of Agricultural and Life Sciences, The University of Tokyo, Tokyo, Japan
e-mail: aokurot@g.ecc.u-tokyo.ac.jp; asami@g.ecc.u-tokyo.ac.jp

M. Ebina
National Agriculture and Food Research Organization (NARO), Institute of Livestock and Grassland Science (NILGS), Tochigi, Japan
e-mail: triticum@affrc.go.jp

T. Nagamine et al. (eds.), *Plant Genetic Resources (PGR) for Sustainable Crop Production*, Sustainable Development Goals Series, https://doi.org/10.1007/978-981-96-7117-5_9

Keywords

Mongolian grassland · Restoration · *Chloris virgata* · Nomad Garden · Medicinal herb · Livestock health

9.1 Introduction

9.1.1 The Mongolian Steppe: Characteristics and Plant Adaptations

The Mongolian steppe exists under extremely unique environmental conditions compared to temperate regions, with short summers and dry and harsh winters. Plants growing in this steppe possess numerous characteristics that suggest their evolution has fostered specialized adaptations, such as growth-promoting functions and stress tolerance, compared to plants in temperate regions. Additionally, numerous plants grown under these extreme conditions have medicinal properties that can promote animal health, which are not found in plants from temperate regions. Knowledge about these plants has been passed down as traditional lore among nomadic people who have maintained civilization in the Mongolian steppe.

9.1.2 Degradation of the Mongolian Steppe and Past Efforts

In recent years, climate change has triggered frequent extreme weather events, desertification, water-resource constraints, soil degradation, and other issues leading to global food insecurity. Unpredictable weather caused by global warming has also affected the Mongolian steppe, leading to its degradation as arid areas expand. Additionally, the number of livestock has increased from 23 million to 66 million over the past 28 years, resulting in "overgrazing" (Fig. 9.1). This has inflicted severe damage on the plant communities of the Mongolian steppe. In particular, frequent grass shortages have occurred during the harsh winters in recent years, forcing many nomadic herders into situations where their herds perish entirely, leading them to abandon their pastoral lifestyles. These displaced herders move to the capital, Ulaanbaatar, swelling the ranks of the urban poor and contributing significantly to the deterioration of the urban environment, which has become a major social issue in Mongolia. To mitigate the effects of overgrazing on grassland degradation, the government has promoted policies encouraging modern nomads to revert to the traditional practice of frequently relocating their portable dwellings, known as ger, every month or less, as opposed to staying in one place for extended periods. However, few nomads follow these recommendations, making it difficult to distribute livestock pressure across grasslands and exacerbating steppe degradation.

9.1.3 New Approaches to Address the Issues

New efforts have been made to leverage the unique traits of native Mongolian steppe plants through advancements in molecular biology, molecular physiology, molecular breeding, natural product chemistry, and chemical biology. These efforts have focused on developing plant cultivation systems based on sustainable agricultural practices to identify optimal conditions for cultivating high-performing plants, thereby enhancing the productivity and value of Mongolian steppe plant resources. The ultimate goal is to restore the degraded steppe ecosystems and revitalize Mongolia's livestock industry. As a novel approach, the recovery of degraded grasslands can be pursued using annual plants. Annual plants are highly effective for immediate restoration in degraded areas. Maintaining the recovered state and supporting the health of nomadic livestock require subsequent sowing of perennial plants and high-performing forage grasses. Ultimately, the aim is to disseminate the resulting plant seeds and cultivation systems throughout the Mongolian steppe to revitalize the nomadic lifestyle and contribute to the broader development of Mongolian society.

Fig. 9.1 The picture of the degraded grasslands targeted for restoration in this project

This project mainly consisted of three interrelated subthemes, each authored by its respective leader(s).

9.2 Characterization and Application of *Chloris virgata*

9.2.1 Identification of *C. virgata* as a Rapid-Growth Plant from Mongolian Grasslands

The summer in the Mongolian grasslands is short, lasting less than two months. During this period, with a small amount of rainfall, the grassland plants germinate and grow rapidly, causing the Mongolian fields to change from brown to green. In southern Mongolia's Gobi region, these extremes intensify, where summer and winter-like conditions can coexist within a single day owing to dramatic temperature fluctuations. Certain plant species have adapted to survive in challenging environments. Under such extreme conditions, it is hypothesized that plants in the Mongolian grasslands have undergone evolutionary selection, developing genetic mutations that enable rapid growth in response to minimal rainfall and survival under severe environmental stress conditions, including drought, high soil salinity, and extreme high or low temperatures.

Dr. Batkhuu of the National University of Mongolia collected seeds from 40 grassland plant species that possessed high growth ability and/or stress resistance based on traditional nomadic knowledge. Subsequent observations of seed germination speed and early seedling growth revealed that the grass species *C. virgata* demonstrated exceptionally rapid germination. Locally known as *Govin Yorkhug* (translated as "Tail of the Gobi Marten"), *C. virgata* is traditionally recognized by Mongolian nomads for its ability to grow in rocky and dry land and quickly recover from livestock grazing. However, fewer than ten scientific studies have been published on this species, highlighting its largely unexplored status in academic and biological research, particularly in developed countries. Detailed analysis revealed that although crops such as rice (*Oryza sativa*) and wheat (*Triticum aestivum*) typically take >48 h to germinate after water absorption, *C. virgata* begins germinating within 5 h of water treatment. Furthermore, this species exhibited more rapid organ elongation and branch formation during the early and intermediate growth stages than rice or wheat (Fig. 9.2). Additionally, *C. virgata* has demonstrated remarkable resistance to environmental stresses such as drought, high salinity, and intense light. These findings suggest that *C. virgata* possesses an exceptional combination of rapid growth and stress resistance, underscoring its vitality and adaptability to severe conditions.

9.2.2 Discovery of *C. virgata* Colony in the Gobi Desert

To observe the growth conditions of *C. virgata* in Mongolian grasslands, we conducted two field explorations in collaboration with Kyoto University and the National University of Mongolia in the southernmost region of the Gobi Desert in May 2021 and July 2024. After departing from the capital city of Ulaanbaatar by car, we reached a stark desert area after three days. This lifeless zone was devoid of visible plants, birds, or insects. We discovered a colony of *C. virgata* 10 min by car from this lifeless zone (Fig. 9.3). The discovery of this surviving colony at the boundary of a lifeless area suggests that *C. virgata* can survive as a pioneer species. This grass demonstrates exceptional adaptability to severe environmental conditions, including extreme drought, high-intensity sunlight, high temperature, and high salinity, at the southernmost edge of the Gobi Desert.

9.2.3 Transcriptome Analysis of *C. virgata* Using Next-Generation Sequencing

We performed a de novo RNA-seq analysis to study the transcriptome derived from the genomic DNA of *C. virgata*. Gene sequencing was performed using Illumina's next-generation sequencer, MiSeq, whereas quantitative analysis of gene expression during the four growth stages (2, 8, 10, and 21 days post-germination) was conducted using Thermo Fisher's Ion Proton

Fig. 9.2 *Chloris virgata* showing rapid growth ability

Fig. 9.3 A great colony of *Chloris virgata* discovered in front of the "lifeless desert zone" in Gobi

platform. In the first stage of the analysis, 21,589 genes were identified in *C. virgata*. Among these, approximately 19,000 genes showed homology with rice (*O. sativa*) genes, and approximately 18,000 genes were homologous to those of the model plant *Arabidopsis thaliana*. This analysis suggests that >10% of the genes in *C. virgata* represent species-specific genetic features.

The second stage of the analysis focused on gene expression levels. During the early germination stage (2 days post-germination), increased expression of genes associated with the biosynthesis and signaling of gibberellins, which are plant hormones that promote germination, was observed. Eight days post-germination, corresponding to the seedling growth activation stage, genes involved in the biosynthesis and signaling of brassinosteroids, which are plant hormones known to promote growth, exhibited elevated expression. These findings suggest that the rapid growth characteristics of *C. virgata* are strongly influenced by the functional genes encoding plant hormones (Bolortuya et al. 2021).

Currently, we are identifying novel genes with high expression from *C. virgata* mRNA during the early growth stages or under environmental stress conditions. These genes were induced in *A. thaliana* genome via genetic transformation to analyze their roles in growth and stress resistance. This study aimed to identify the genes that may enhance agricultural productivity. In the future, it is anticipated that genes from Mongolian grass-derived *C. virgata* will contribute to improving crop productivity worldwide.

9.2.4 An Evaluation of the Nutritional Value and Yield

Evaluation of the nutritional value and yield of Mongolian grassland plants demonstrating high biomass productivity under nutrient-poor conditions revealed that *C. virgata* achieved yields nearly equivalent to those of related forage species grown in tropical and subtropical regions under sufficient nutrient conditions. Additionally, it exhibited approximately double the yield of the Japanese accession "KOBE." These results confirm the superior growth characteristics of Mongolian *C. virgata* under nutrient-poor conditions. A common cultivation method applicable to Mongolia and Japan was examined, and methods for evaluating yield and nutritional value were established. The preliminary variety registration was completed in Mongolia.

9.2.5 The Necessity of Establishing *C. virgata* Varieties

In a study on male adult sheep during fattening under various feeding regimes, an average dry matter intake of approximately 1.2 kg/day and an energy intake of approximately 2.9 Mcal/day were observed, resulting in a weight gain of 200 g/day (Ma et al. 2019). Additionally, Furuto et al. (2018) estimated the energy consumption of pregnant ewes on the Mongolian steppe in winter to be 6.23 Mcal/day (1.4 kg of total digestible nutrient [TDN]); however, the estimated energy intake from feeding behavior was only 0.84 Mcal/day, resulting in a negative energy balance of 5.39 Mcal/day. Furthermore, Uehara et al. (2015) estimated the forage intake of sheep in the Mongolian grasslands from September 2012 to April 2013. They reported that the dry matter intake was 1.73 kg in October and 1.41 kg in February. By April, the body weight of the sheep was 88.7% of their body weight in October.

From these cases, it can be considered that an energy intake of approximately 3.0 Mcal/day (0.68 kg TDN/day) is necessary without considering the environmental load energy. In our study, the TDN content (%) of *C. virgata* was approximately 50%; therefore, the required energy can be met with a dry matter intake of 1.36 kg/day. The annual dry matter yield per 10a was 139 kg/10a in Urgun and 95 kg/10a in Uulbayan. If the annual consumption is stabilized, the total dry matter requirement per pregnant ewe is 496.4 kg, and to obtain the necessary dry matter yield solely from *C. virgata*, an area of 36a/head and 50a/head would be required. Mongolia has a land area of 25 million ha; therefore, if it is possible to satisfy one pregnant ewe per 36a grazing capacity as in Urgun, it would be possible to raise the current 65 million sheep.

Therefore, the *C. virgata* we focused on has sufficient grazing capacity to restore grasslands and is a valuable grass species for registration and dissemination in Mongolia owing to its rapid initial growth and ecologically important role as a pioneer plant.

9.3 Medicinal Herbs in Mongolian Grassland for Livestock Health

9.3.1 Pasture Grasses and Shrub as Medicinal Plants on the Mongolian Plateau

In Mongolia, due to the characteristics of nomad livestock, it is more common for them to consume plants that are native to the site at the time of the season rather than eating specific forages and pastures that have been artificially prepared and planned for them. Mongolian nomads also recognize that all plants growing locally are pasture grasses that can be used in some way for animals and their lives.

Naturally, many plants are thought to have negative effects due to their toxicity; however, by skillfully selecting the time, season, and method of use, animals and nomads can use plants that are generally considered poisonous.

We have continued to analyze the medicinal properties of various plants, drawing on knowledge and traditions about the usefulness of Mongolian plant resources, with the intention of converting knowledge into components that could become medicinal seeds and into the communication between organisms through plant components (Murata and Batkhuu 2021).

9.3.2 Yargui: The Project that Began with a Tonic Flower

In the conception of the SATREPS project, our Output 2 Medicinal Plants Group aimed to identify useful plants and their constituents by deciphering the characteristics of native plants through animal interactions and stories from local people. “Yargui” is a symbolic plant species that has garnered interest. “Yargui,” a plant of the genus *Pulsatilla* (Ranunculaceae), is known to the Mongolian people. Its flowers bloom in early spring before other plants and are welcomed by livestock and nomads after a hard winter as a tonic that, if eaten, helps them stay healthy throughout the year.

Specifically, we focused on purple-flowered *Pulsatilla bungeana* and yellow-flowered *P. flavescens*, which are found in our research area, and explored the constituents of their flowers (Ganchimeg et al. 2019). This was followed by a survey of how many livestock animals eat these plants. Together with Mongolian researchers, they tried to determine the effects of feeding a large number of flowers only to goats. At present, although issues remain in terms of plant resources and analysis of medicinal effects, it is positioned as one of the promising targets for future research.

9.3.3 Local Medicinal Herbs Significantly Used by Mongolian Veterinarians

Veterinarians in Mongolia use local herbs to treat their livestock. *Woodsia ilvensis* is a medicinal fern that is used to treat animal trauma. Because *W. ilvensis* is used to treat wounds, we focused on plasmin, which enhances the fibrinolytic system and is involved in bleeding and inflammation. We found that a water-soluble fraction extract of this plant had significant plasmin inhibitory activity (Otgonsugar et al. 2023). Furthermore, our finding that plasmin inhibitory activity was also observed in some isolated and purified compounds can be considered one achievement. However, this is insufficient to explain the strong activity of these extracts. We also believed that effective points other than plasmin should be investigated, and further research is required. This is an interesting research subject from the perspective of the plant’s actual use by the Mongolian people.

9.3.4 Proposal for Livestock Supplements Using Local Useful Plants

We are searching for materials, including *Artemisia sieversiana* and *Rheum undulatum*, that could contribute to the health of livestock. Currently, we are focusing on *Artemisia species*, which are thriving plants in the area; however, this diversity is being lost due to overgrazing. Currently, we are examining their effects on nomadic livestock animals from an ingredient standpoint and pursuing pharmacological aspects with the goal of applying them as livestock supplements.

The characteristic components of *A. sieversiana* are lignans, flavonoids, and terpenoids (Nurbek et al. 2020). This study focused on *A. sieversiana* as a valuable pasture for livestock animals in early spring and fall and how it affects the animals that consume it. In addition, studies on the constituents of *R. undulatum* have shown that anthraquinones are the major constituents. It is thought to affect the gastrointestinal system; if appropriate use is proposed, this plant may be able to support digestion and absorption systems in livestock. Collaborative research with the Institute of Veterinary Medicine of Mongolia has attempted to identify Mongolian patents and list them in the Livestock Pharmacopoeia of Mongolia. Additionally, commercialization of the two plants mentioned above as a livestock supplement will be one of the implementation parts of this project.

9.3.5 Development of Mongolia-Japan Cooperative Relations Through Collaborative Research

We have obtained characteristic flavonoids from *Scutellaria scordifolia* which are closely related to *S. baicalensis*, a crude drug used in Kampo medicine, and are disseminating research results while exploring their potential applications (Nurbyek et al. 2023). Using local resources, Japan-Mongolia joint research develops projects utilizing the strengths of both parties. Currently, pasture research and the application of its results are ongoing, focusing on Asteraceae, Fabaceae, and Rosaceae plants on the Mongolian steppe, which are key to their position in the current ecosystem and their relationship with livestock. The identification of materials from Mongolian pasture grasses is particularly useful for maintaining animal health and proposing effective methods for their use.

9.4 Efforts Toward the Development and Dissemination of Cultivation Techniques for Useful Plants

9.4.1 Diagnosis, Treatment, and Prevention of Desertification and Grassland Degradation

Various approaches to combating desertification are often compared with medical practices. These include "diagnosis," which assesses the extent of desertification and environmental carrying capacity; "treatment," which involves restoring degraded land through measures such greening; and "prevention," which focuses on land use and management in proactive ways that prevent desertification. In this chapter, we introduce efforts related to the development of field cultivation and introduction techniques and the dissemination and implementation of results from the perspective of diagnosing, treating, and preventing desertification and grassland degradation, which are the main goals of the SATREPS project.

9.4.2 Diagnosis of Grassland and Growth Conditions for Candidate Plants

Environmental gradients are clearly formed in Mongolia, ranging from humid and cold in the north to dry and warm in the south. In response to these gradients, various types of grassland vegetation, from forest steppes in the north to typical and desert steppes, culminate in southern desert regions. Additionally, depending on the grazing intensity, there are various types of grasslands, ranging from highly productive, high-quality grasslands to degraded grasslands with low biomass and fodder values. Therefore, to introduce useful plants, it is crucial to first diagnose the condition of the grassland, set appropriate target vegetation, and select the most suitable plant species for introduction.

In degraded grasslands, the introduction of pioneer species that are resistant to stress and disturbance is considered the best solution for restoration. In the present study, *C. virgata*, an annual grass of the Poaceae family with rapid growth and high environmental stress resistance, was initially selected as a candidate plant. However, where does this species grow in Mongolia? To answer this question, we conducted a distribution survey of *C. virgata* in central and southeastern Mongolia. The results revealed that large colonies of *C. virgata* were primarily distributed along the edges of playa-type lowlands, and their growth was significantly influenced by annual and seasonal rainfall fluctuations. Additionally, the distribution of this species was restricted to sites with high disturbance and its dominance rapidly decreased as grassland vegetation developed and disturbance decreased. This confirms its suitability as a pioneer plant. Furthermore, dead plants remain throughout winter and spring, making them a valuable fodder for livestock. Recent research has shown that dead plants contribute to the suppression of dust emissions during spring. Thus, *C. virgata* is a promising species that offers three benefits: the restoration of degraded grasslands, provision of livestock feed, and suppression of dust emissions.

9.4.3 Development of Cultivation Techniques for Candidate Plants as a Treatment

Next, we focused on developing cultivation techniques to introduce candidate plants into fields to restore (treat) degraded grasslands. We established three model test sites in Töv Province in typical grassland areas, with varying soil conditions and degrees of degradation (Batsumber,

Arkhust, and Hustai), and conducted field trials of *C. virgata* to investigate the conditions required for its cultivation. Through 3 years of field trials and indoor germination tests, we determined the optimal conditions for sowing depth, sowing interval, sowing time, and germination temperature. Additionally, we examined measures to protect plants from weather-related disasters, such as hail and heavy rain, and damage from pests, such as rodents and locusts. For example, we found that spreading sowing times could reduce the damage from weather-related disasters from sowing to the early growth phase.

Once the recovery effects of introducing pioneer plants, such as *C. virgata*, are confirmed, the next step is to introduce stabilizing plant species that help further stabilize the environment, such as perennial grasses, which are the dominant species in Mongolia's native grasslands. At the Hustai test site, we conducted cultivation trials with perennial grasses (*Stipa sibirica*, *S. baicalensis*, and *S. krylovii*) as stabilizing plant candidates and examined the conditions for their cultivation. The results confirmed that, with appropriate pest control, these species can be successfully introduced and cultivated in the field.

In Mongolia's harsh environment (dry, hot, and saline with strong sunlight and wind), it is difficult to establish plants using conventional sowing techniques alone. Therefore, this project focused on developing supporting technologies to promote germination and establishment, including seed improvement, soil improvement, and sowing method optimization.

As an example of seed improvement, we developed a seed-coating method in which seeds were coated with fertilizers and water-retaining components to form pellets. We tested this method using *Chloris* seeds and developed a protocol to produce stable pellet seeds using diatomaceous earth as a powder and hydroxyethyl cellulose as a binder. Based on field trials at the model test sites, we found that the germination rate of *C. virgata* seeds in the field was approximately 30–40% and adjusted the pellet size to ensure that each pellet contained at least three seeds (2.0–4.0 mm in diameter). We also tested coating materials (powders and binders) to enhance germination and seedling growth and found that mixing activated charcoal and urea significantly improved germination and seedling growth. The effectiveness of the pellet-coated seeds was confirmed through field trials, in which improved germination rates were observed. We have been developing other seed processing technologies (such as coating, pelletizing, and awn removal), as well as soil chemical and physical improvements (such as organic material application, no-till cultivation, and crusting) and sowing methods (mixed sowing to enhance positive plant-plant interaction, called facilitation).

9.5 Toward the Establishment of the "Mongolian Nomad Garden" as an Information Hub for Prevention

We aimed to unravel the traditional knowledge of nomadic people regarding useful plants through cutting-edge science and applying it to grassland restoration. However, as we expanded our broad-scale interview surveys on the use of forage and medicinal plants, it became evident that the traditional knowledge passed down among nomads was not always sufficiently shared within their communities. This realization highlights the growing need to simultaneously enhance mechanisms to promote the dissemination and education of such knowledge.

In response, we decided to establish a new demonstration site in Arkhust to complement the dissemination of cultivation techniques while further strengthening capacity building and educational effects for herders. Specifically, we constructed a test site for sustainable small-scale cultivation systems that combined forage cultivation near winter camps with medicinal plant cultivation and windbreak planting. In addition, we built a demonstration site to provide information on the functions of useful plants and related cultivation techniques. This site is referred to as the "Mongolian Nomad Garden."

Japan and Mongolia are collaboratively advancing the development of the "Nomad Garden" to serve as a hub for disseminating diverse knowledge and preventive techniques related to plant conservation and utilization.

9.6 Afterword

To date, there has been limited physiological and pharmacological research on the usefulness of Mongolian grassland plants for livestock. This project focuses on this aspect and aims to enhance the value of degraded Mongolian grasslands based on foundational research findings. At this point, basic biological and chemical knowledge regarding grassland plants was obtained, and the feasibility of implementation confirmed, indicating that the project achieved its intended outcomes.

This project aimed to apply insights gained from genetic analyses and studies on active substances to increase the production of Mongolian pasture plants and maintain livestock health. Simultaneously, genes that accelerate plant growth rates, enhance plant biomass production, or produce active compounds (targeted for isolation from Mongolian grassland plants) are expected to present new research opportunities in plant molecular biology, which has primarily focused on temperate plants. This can significantly contribute to the advancement of science and technology.

We hope that the outcomes of this project will contribute to the future of Mongolian nomadic pastoralism.

References

Bolortuya B, Kawabata S, Yamagami A, Davaapurev B, Takahashi F, Inoue K, Kanatani A, Mochida K, Kumazawa M, Ifuku K, Jigjidsuren S, Battogtokh T, Udval G, Shinozaki K, Asami T, Batkhuu J, Nakano T (2021) Transcriptome analysis of *Chloris virgata*, which shows the fastest germination and growth in the major Mongolian grassland plant. Front Plant Sci 28:684987. https://doi.org/10.3389/fpls.2021.684987

Furuto S, Luodian, Umemura K, Yamanaka N, Shinoda M, Yoshihara Y (2018) Energy balance, and nutritional status of mongolian grazing sheep in winter. https://drylands.has.env.nagoya-u.ac.jp/shinoda_lab/doc/201803mieuniv.pdf

Ganchimeg D, Batbold B, Murata T, Davaapurev B, Munkhjargal T, Tuvshintulga B, Suganuma K, Igarashi I, Buyankhishig B, Sasaki K, Batsuren D, Batkhuu J (2019) Flavonoids isolated from the flowers of *Pulsatilla flavescens* and their anti-piroplasm activity. J Nat Med 73:633–640. https://doi.org/10.1007/s11418-019-01294-8

Ma T, Wan F, Yang D, Deng, Yang K, Diao Q (2019) Growth performance, nutrient digestibility, and slaughter traits of male fattening lambs under different feeding standards. Anim Nutr 5(1):74–79. https://doi.org/10.1016/j.aninu.2018.07.002

Murata T, Batkhuu J (2021) Biological activity evaluations of chemical constituents derived from Mongolian medicinal forage plants and their applications in combating infectious diseases and addressing health problems in humans and livestock. J Nat Med 75:729–740. https://doi.org/10.1007/s11418-021-01529-7

Nurbek S, Murata T, Suganuma K, Ishikawa Y, Buyankhishig B, Kikuchi T, Byambajav T, Davaapurev B, Sasaki K, Batkhuu J (2020) Isolation and evaluation of trypanocidal activity of sesquiterpenoids, flavonoids, and lignans in *Artemisia sieversiana* collected in Mongolia. J Nat Med 74:750–757. https://doi.org/10.1007/s11418-020-01429-2

Nurbyek S, Buyankhishig B, Suganuma K, Ishikawa Y, Kutsuma M, Abe M, Sasaki K, Davaapurev B, Batkhuu J, Murata T (2023) Phytochemical investigation of *Scutellaria scordiifolia* and its trypanocidal activity. Phytochemistry 209:113615. https://doi.org/10.1016/j.phytochem.2023.113615

Otgonsugar P, Buyankhishig B, Undrakhbayar T, Bilguun B, Sasaki K, Davaapurev B, Batkhuu J, Byambajav T, Murata T (2023) Phytochemical investigation of aerial parts of *Woodsia ilvensis* and its plasmin-inhibitory activity *in vitro*. Phytochemistry 215:113826. https://doi.org/10.1016/j.phytochem.2023.113826

Uehara A, Ayush E, Ganbaatar O, Yamasaki S, Shindo K (2015) Forage intake estimation of grazing sheep by lignin method during late summer and the following spring in steppe zone in Mongolia. J Arid Land Stud 25(2):17–24. https://doi.org/10.14976/jals.25.2_17

10 Prospects for Sustainable Cassava Production and Distribution in Cambodia and Vietnam

Tamon Baba, Hisako Nomura, Sovannara Moun, Tha Than, Nguy Xuan Truc Nguyen, Nien Chau Nguyen, Pao Srean, and Kasumi Ito

Abstract

Cassava is an important cash crop in Southeast Asia, especially in Thailand, Cambodia, and Vietnam. However, it is susceptible to viral diseases and insect pests. To secure cassava yield and ensure profit from cassava production and distribution in Cambodia and Vietnam, our research team implemented a sustainable cassava seed production and distribution system in Cambodia and Vietnam.

Keywords

Cassava · Cash crop · Cambodia · Vietnam · Cassava mosaic disease · Invasive pest management

10.1 Introduction

Cassava (*Manihot esculenta* Crantz) is widely cultivated in more than 90 countries, mainly in the tropical and subtropical latitudes of Latin America, Africa, and Asia (Fig. 10.1). Its tuberous root is an important source of starch as a staple food; it is also used as a biofuel, an ingredient for cosmetics and animal food, and a raw material for the production of paper, textiles, and glue. Hence, cassava is an important source of income for both cassava producers and national governments. As the demand for cassava starch has sharply increased in recent years, cassava has become one of the most important cash crops, especially in Southeast Asia; in particular, 34,068,005 tons of cassava were produced in Thailand, 17,698,784 tons in Cambodia, and 10,626,862 tons in Vietnam in 2022, making them the third-, fifth-, and eighth-largest producers of cassava worldwide, respectively (FAO 2024).

However, cassava production in these countries faces several challenges. As cassava is propagated by planting the cuttings of the remaining

T. Baba (✉)
College of Letters, Ritsumeikan University, Kyoto, Japan
e-mail: babatmn@fc.ritsumei.ac.jp

H. Nomura
Faculty of Agriculture, Kyushu University, Fukuoka, Japan

S. Moun · T. Than · P. Srean
Faculty of Agriculture and Food Processing, National University of Battambang, Battambang, Cambodia

N. X. T. Nguyen
Ochna Vietnam, JSC, Hanoi, Vietnam

N. C. Nguyen
Faculty of Agronomy, Nong Lam University, Ho Chi Minh, Vietnam

K. Ito
International Center for Research and Education in Agriculture, Nagoya University, Nagoya, Japan

T. Nagamine et al. (eds.), *Plant Genetic Resources (PGR) for Sustainable Crop Production*, Sustainable Development Goals Series, https://doi.org/10.1007/978-981-96-7117-5_10

Fig. 10.1 Cassava field in Trang Bom district, Dong Nai province, Vietnam, October 2016. (Photo by T. Baba)

stems after its tuber is harvested, it is highly susceptible to viral diseases, such as witches' broom (Alvarez et al. 2013) and cassava mosaic disease (CMD), and harmful pests, including mealybugs, red mites, and scale insects. Notably, since the outbreak of Sri Lankan CMD (SLCMD) was reported in Cambodia in 2015 (Wang et al. 2015) and then in Vietnam in 2018 (Uke et al. 2018), CMD, which is caused by cassava mosaic virus (CMV), has spread rapidly in Southeast Asia (Tokunaga et al. 2018). Whiteflies, which serve as the vector of CMD, and anthropogenic activities, such as transporting infected stems across countries (Delaquis et al. 2018) and practicing informal and self-regulated seed distribution, help spread the disease.

A government-led extension system is necessary for sustainable cassava production (Tokunaga et al. 2018). In Cambodia, the Department of Agricultural Extension, under the Ministry of Agriculture, Forestry and Fisheries, was established in 1995. In Vietnam, a public extension system was founded in 1993. Furthermore, the Hung Loc Agricultural Research Center (HLARC) in Dong Nai, a province in Vietnam, has produced and distributed several cassava varieties. However, the shortage and poor quality of human resources, funding, techniques, extension materials, and systems were considered the main issues of these systems at that time; fortunately, the situation has improved recently (see Conclusion).

In this context, we executed a project titled "Development and Dissemination of a Sustainable Production System Based on Invasive Pest Management of Cassava in Vietnam, Cambodia, and Thailand" (2016–2021), which aimed to establish a sustainable and healthy cassava seed production and dissemination system with a management method for invasive cassava diseases and insect pests in Cambodia and Vietnam by applying the cassava cultivation techniques of Thailand, which pioneered cassava cultivation, to overcome the above-mentioned challenges. The National University of Battambang, Battambang City, Cambodia (NUBB), HLARC, and Rayong Field Crops Research Center, Rayong Province, Thailand (RFCRC), were the participating research institutions in counterpart country.

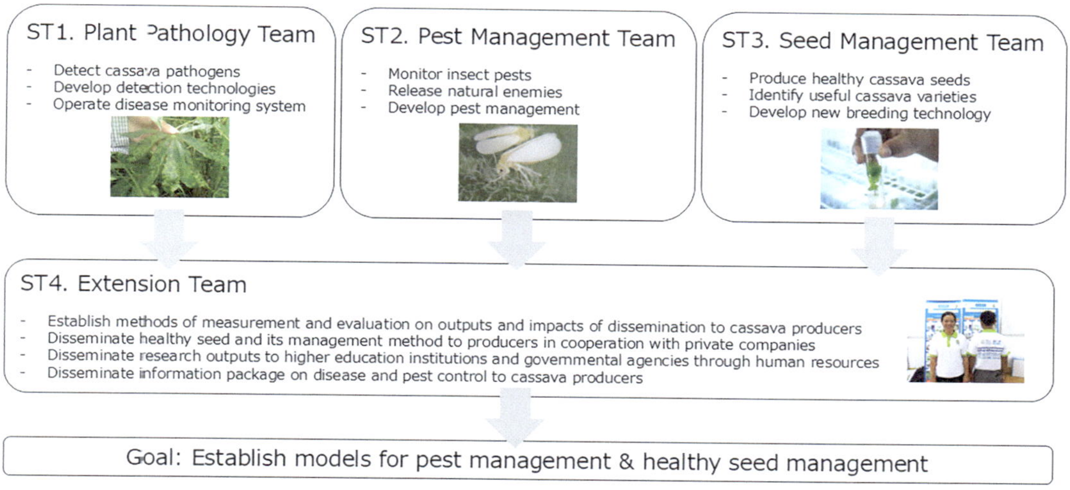

Fig. 10.2 Project outline. (The illustration was created by the authors)

The international team collaborated and co-benefitted from performing the following activities: the Plant Pathology Team detected cassava pathogens, developed detection technologies, and operated a disease monitoring system; the Pest Management Team monitored insect pests, examined ways to release natural enemies, and developed pest management strategies; the Seed Management Team produced healthy cassava seeds using different methods, identified useful cassava varieties, and developed new breeding technologies; and the Extension Team worked together with other teams by utilizing the knowledge accumulated and implemented to establish methods of output measurement and evaluation of and the impacts of information dissemination to cassava producers, disseminate healthy seeds and management methods to producers in cooperation with private companies, disseminate research outputs to higher education institutions and governmental agencies through human resources, and disseminate information packages on disease and pest control to cassava producers. Then, finally, the project established a model for pest and healthy seed management.

This chapter focuses on the research activities and outputs of the Extension Team to present the prospects for sustainable cassava production and distribution in Cambodia and Vietnam. To establish a sustainable extension scheme including public research institutions, producers, and the private sector that will all benefit from disseminating the technology package developed in this project, we achieved the following four outputs:

Output 1. Establish methods for the measurement and evaluation of outputs and impacts of information dissemination to cassava producers.

Output 2. Disseminate healthy seeds and their management methods to producers in cooperation with private companies.

Output 3. Disseminate research output to higher education institutions and governmental agencies through human resources.

Output 4. Disseminate information packages on disease and pest control to cassava producers.

We will explain each output in detail (Fig 10.2).

10.2 Established Methods of Measurement and Evaluation of Outputs and Impacts of Information Dissemination to Cassava Producers

10.2.1 Survey Area and Sampling Procedure

To establish measurement and evaluation methods, cassava producers in Cambodia and Vietnam were surveyed in 2016 and 2017 to determine the current cassava production status and socioeconomic situation of cassava producers in both countries and identify potential healthy seed producers who would purchase healthy seeds from NUBB or HLARC, multiply them, and establish a healthy seed production business. The identification of the producers' cassava seed preference, distribution, propagation methods, and demand for healthy cassava seeds was an essential first step in the project.

We determined the research target area near the stock seed production fields: NUBB in Cambodia and HLARC in Vietnam. In Cambodia, we selected two major cassava-producing provinces located in the northwestern part of the country that both border Thailand: Battambang and Pailin (Fig. 10.3). First, 24 cassava producers participated in semi-structured interviews, where they were asked about their cultivation methods, annual schedule, and sales destinations in Cambodia. Additionally, owners of six collection and processing points (known as silos in Cambodia) that collect cassava roots, manufacture dried chips, and sell those products either domestically or internationally were interviewed about their customers and sales destinations to identify their distribution channel. The survey was conducted in Battambang and Pailin from July 2017 to February 2018. Second, we employed a structured questionnaire to interview producers about the 2016–2017 cassava production cycle in two districts of Pailin Province and six districts of Battambang Province from April to November 2017. A stratified random sample based on the number of cassava planted was referred to in selecting the respondents. A total of 205 cassava producers participated in the interview: 144 from Battambang and 61 from Pailin.

Meanwhile, in Vietnam, we selected three major cassava-producing provinces in southern Vietnam: Dong Nai, Gia Lai (borders Cambodia), and Tay Ninh (Fig. 10.4). A series of semi-structured interviews based on a questionnaire were conducted from April to December 2017. A total of 182 cassava producers participated in the interviews: 60 from Dong Nai, 61 from Gia Lai, and 61 from Tay Ninh.

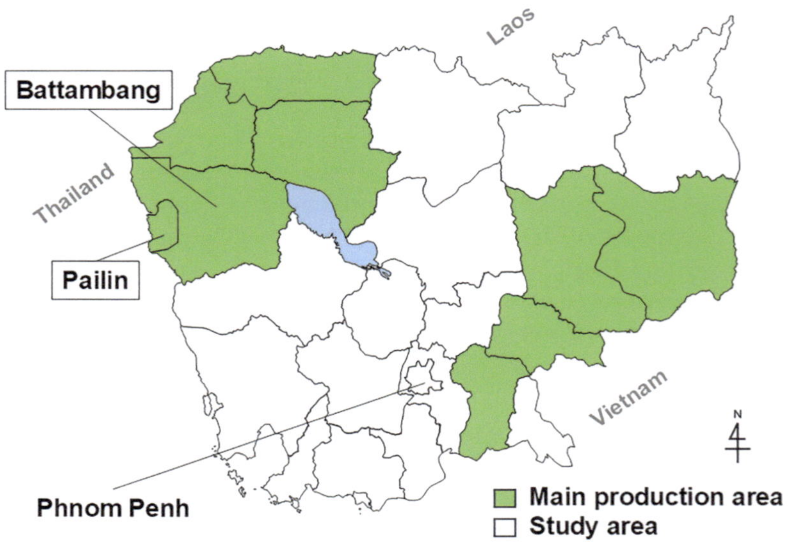

Fig. 10.3 Study area in Cambodia. (The map was created by the authors)

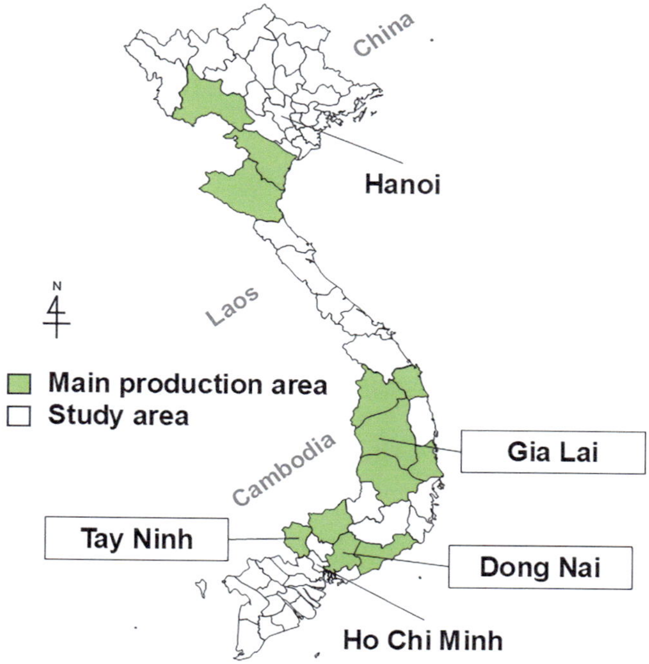

Fig. 10.4 Study area in Vietnam (Nguyen et al. 2023)

10.2.2 Possibility of Producing and Distributing Healthy Cassava Seeds

The interview results (Ito et al. 2021) revealed three different types of distribution channels from producers to consolidating points. The first type is direct distribution, wherein all related processes, such as planting, harvesting, transporting, and selling, are handled by producers. In the second type, except for cultivation, everything is handled by middlemen. The third type featured a combination of cultivators and middlemen. The distribution type also depends on the type of root being traded and the distance from the farm to the collection and processing points. We also found that large-scale cassava cultivators, who operate root collection and processing points, play important roles as collectors, middlemen, and transporters in the cassava root distribution system in Cambodia (Fig. 10.5).

Furthermore, the current status of cassava seeds in Cambodia, including producers' preference for cassava varieties and their demand for healthy cassava seeds, was surveyed (Baba et al. 2021). The interview results revealed that cassava producers in Cambodia can accept and produce healthy cassava seeds. Although they highly consider the starch content of cassava varieties, they do not have a very strong preference for a specific variety and can accept any variants distributed in the market.

In Vietnam, we captured the potential for healthy cassava seed production (Nguyen et al. 2021). The interview results revealed that producers tend to prefer the cassava varieties that are predominant in their communities.

These findings suggest that establishing an efficient and viable distribution mechanism and strategy by upgrading the existing system as its foundation will be one of the most important challenges for making cassava a major industrial crop and increasing revenue from cassava cultivation in Cambodia and Vietnam; nonetheless, cassava variety is an important criterion for cassava producers in Vietnam. While SLCMD was spreading at the time when the survey was conducted, there was high demand for a healthy seed market in both countries. Our market-based healthy seed production system can contribute to

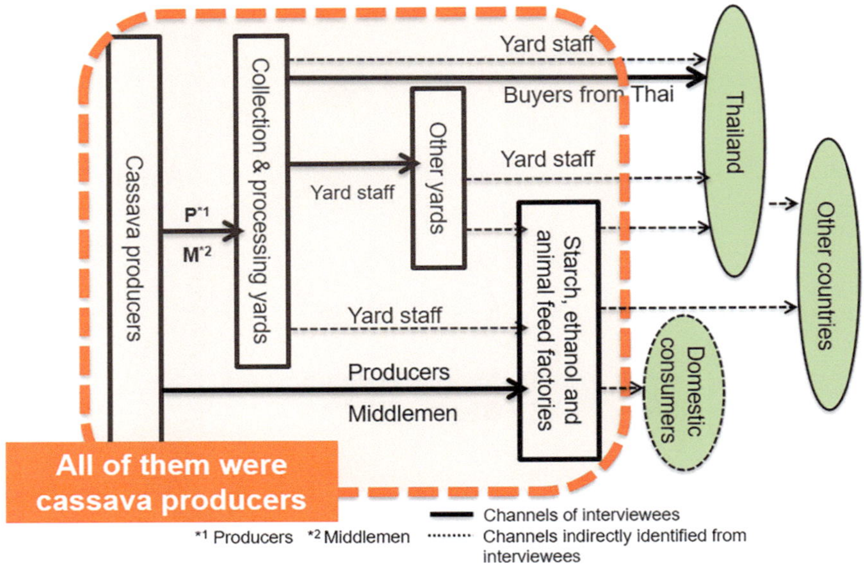

Fig. 10.5 *Cassava root* distribution system in Cambodia (Ito et al. 2021)

sustainable cassava production to help producers increase their income and improve the economic growth of a province.

10.2.3 Socioeconomic Characteristics of Cassava Production

Research on the socioeconomic characteristics of cassava producers is limited in Cambodia and Vietnam. Determining the socioeconomic status of cassava producers, which encompasses land size, income and its main source, and purpose for producing agricultural products, is instrumental to creating a workable extension scheme.

The determinants that influence the "yield" of cassava producers in Cambodia were investigated using a stochastic frontier model (Baba et al. 2022), which revealed that the tractor-or-truck hire cost for transporting cassava tubers is a significantly positive variable, and the uprooting machine-hire cost for cassava tubers was negatively significant. Furthermore, the estimated technical efficiency score of cassava yield was 0.62, indicating that there are several opportunities to increase its technical efficiency. Efficient production methods, including calculating and saving the expenses for cassava yield, have not yet been established in Cambodia because cassava production is relatively new in Cambodia. One of the strategies that cassava producers employ is hiring or purchasing uprooting machines at lower costs. Provincial or regional financial support can help reduce the negative impact of uprooting machines on cassava yield.

The determinants that influence the "profitability" of cassava producers in Cambodia were also investigated (Baba et al. 2023). The findings of this study revealed that fertilizer and labor costs had a significant negative impact on profit across all three models we developed. While commodity prices decreased due to changes in the global trade demand, the local labor market did not immediately adjust. As a result, labor hire costs affected profitability, leading to reduced profit. To counteract the diminishing profit related to producers' strategies, we suggest that cassava producers should refrain from purchasing additional bunches of seeds for replanting; rather, they should plant cassava seeds at optimized density to maximize their profit.

Furthermore, we investigated the income diversity of cassava producers and their roles in sustainable cassava production in Cambodia (unpublished data). The survey results indicate that, on an annual average, the total household income was $2,508.4, of which $1,058.3 (42.19%) was from agricultural income. The average income from cassava production was $462.9, accounting for 18.45% of the total household income and 43.74% of agricultural income. Thus, cassava producers in Cambodia rely on multiple income sources, including both agricultural and non-agricultural activities; cassava production is only one of their several income sources. This diversification of income suggests that they can sustain their livelihoods by engaging in various economic opportunities; however, they may not be full-time producers and face challenges in dedicating their full attention to cassava production.

The household income of cassava producers in Vietnam was also explored (Nguyen et al. 2023). The results showed that, on an annual average, the total household income was $7,071.29, of which $3,648.41 (51.59%) was from agricultural and $3,422.88 (48.41%) from non-agricultural sources. The average income from cassava production was $859.81, accounting for 12.16% of the total household income and 23.57% of agricultural income. Thus, similar to Cambodian cassava producers, Vietnamese cassava producers rely on multiple income sources and do not rely excessively on cassava production.

To our knowledge, this is the first report on the socioeconomic characteristics of small Cambodian and Vietnamese cassava producers. There are various difficulties in cassava yield and profitability, and cassava production is only one source of income diversity. Thus, the model created by our project is needed for the local economy and cassava producers.

10.3 Disseminate Healthy Seed and Its Management Method to Producers in Cooperation with Private Companies

Through the survey above, we identified several potential healthy seed producers in Cambodia and Vietnam who were interested in healthy seed production as a business and could follow the pest management monitoring method and certification system established by the project. This method includes how often monitoring should be carried out, how to select samples to check, how to look at insect density, etc. Healthy seed producers should follow this trend, keep monthly records, and send them to us. When they did the monitoring well, we issued them a certificate so they could sell their seeds labeled as "healthy" to other producers. It should be noted that the trained staff contacted and visited healthy seed producers to train them on how to monitor their cassava fields to prevent SLCMD and support their production.

The stock seed production fields in the NUBB and HLARC began distributing healthy cassava seeds in April 2018. Thereafter, we started to provide seeds to potentially healthy seed producers every year. While some of them have stopped this business after encountering several difficulties, such as floods in their fields, and some newcomers became interested in this business, 29 healthy seed producers in both countries (16 in Cambodia and 13 in Vietnam) tried to produce healthy seeds until 2022. Some sold healthy seeds to general producers and succeeded in making profit.

Furthermore, during 2016 and 2017, we surveyed existing extension systems in Cambodia, Vietnam, and Thailand and visited several institutions, such as the provincial departments of Agriculture, Forestry, and Fishery in Cambodia and RFCYC in Thailand. We found that there were not enough government-led extension systems in Cambodia and Vietnam at that time, whereas Thailand had ample experience in disseminating technologies in cassava, not only in agricultural practices but also in pest control. Thus, we created an information package based on leaflets and posters from the RFCRC, in addition to the outputs from other STs and knowledge of HLARC in 2018 (Figs. 10.6, 10.7, and 10.8). These were used to train the aforementioned healthy seed producers, in addition to the others explained in the following chapters.

Fig. 10.6 Poster "Guideline for Prevention and Treatment of Cassava Mosaic Disease in Cassava" (Vietnamese version). (Created by the project)

10.4 Disseminate Research Outputs to Higher Education Institutions and Government Agencies Through Human Resources

We disseminated the above-mentioned information package, together with outputs from other STs, to researchers, students, and officials. Through joint research, in addition to the survey method, nine researchers from Cambodia, Vietnam, and Thailand acquired the necessary knowledge and skills regarding healthy cassava seeds and their production. Four of them are authors of academic journal articles. We conducted training workshops and distributed leaflets and posters to local government officials and general cassava producers in Cambodia and Vietnam in 2019 and 2021. A total of 105 officials participated in the workshops, and 354 officials received information packages from the government and cooperated with international organizations. Furthermore, we created two videos summarizing the importance of healthy seed production and monitoring methods and then distributed them online. One workshop was broadcasted by the CNC and CTN news channels in Cambodia (Fig. 10.9).

10.5 Disseminate Information Packages on Disease and Pest Control to Cassava Producers

Information packages on disease and pest control, in the form of posters or leaflets, were disseminated to a total of 1,460 cassava producers who participated in several workshops we conducted in Cambodia and Vietnam. After the workshops had ended, we visited several produc-

Fig. 10.7 Leaflet "Guidelines for Prevention and Treatment of Cassava Mosaic Disease in Casava" (Khmer version). (Created by the project)

Fig. 10.8 Leaflet "Technical guidelines for sustainable cassava cultivation in Southern Vietnam provinces" (in Vietnamese). (Created by HLARC)

ers and found that our posters were displayed on house walls.

To evaluate the effectiveness of the dissemination of information packages to cassava producers, we assessed their knowledge, attitude, and practice (KAP) levels regarding cassava pests and diseases before and after the information dissemination campaign. Similarly, to assess the effectiveness of our extension scheme of healthy cassava seed, we determined cassava producers' recognition of the added value of healthy seeds by their willingness to pay (WTP) for healthy seeds without infection certifications after they had participated in knowledge transfer interventions. Based on a randomized controlled trial design, we conducted the first survey, interventions by disseminating posters or workshops and posters, and the second survey. The selected participants in the first survey were randomly assigned to three equal groups: "treatment 1," "treatment 2," and "control." The "treatment 1" group only received posters, the "treatment 2" group received posters and attended workshops, and the "control" group did not receive any sensitization material. We then analyzed the changes in their KAP scores before and after the intervention. After the intervention, we conducted the second survey and compared the changes in KAP (Fig. 10.10).

In Cambodia, the first survey was conducted in October and November 2019 (Moun et al. 2021, 2024). A total of 468 cassava producers were randomly selected, assigned to three equal groups, and received the intervention. Then, the second survey was conducted from November 2019 to February 2020. The survey results revealed that the workshops and poster are effective methods for improving the KAP of cassava producers in Cambodia; however, their WTP for healthy seeds did not significantly change even after the workshop and poster investigation.

In Vietnam (unpublished data), the first survey was conducted in July and August 2018, and 312 cassava producers were randomly selected for an

Fig. 10.9 CTN news broadcasting a workshop in Cambodia

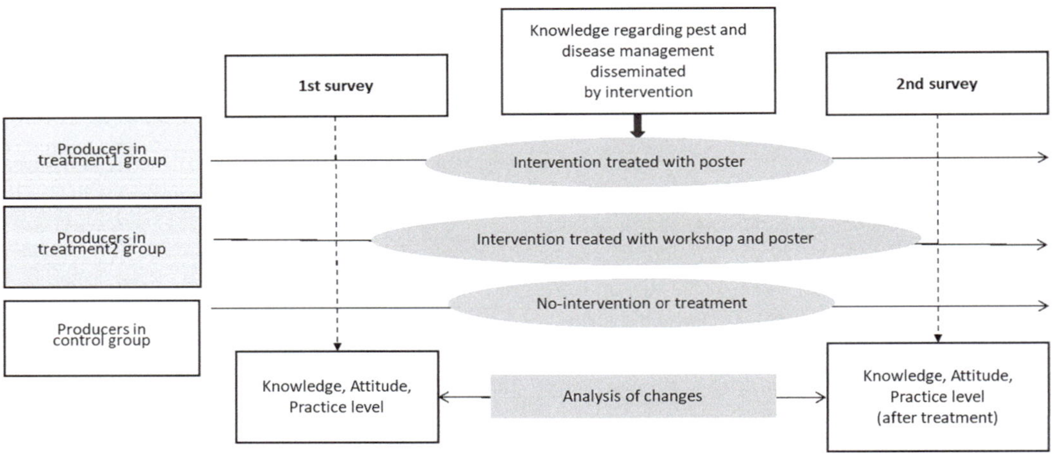

Fig. 10.10 Randomized controlled trial design for knowledge transfer. (Created by the authors)

interview and data collection. The participants were randomly assigned to three equal groups and received the intervention. Subsequently, the second survey was conducted in January and February 2019. The survey results revealed that both "treatment 1" and "treatment 2" groups enhanced their KAP levels after the intervention; however, their WTP for healthy seeds tended to decrease. Subsequently, their confidence in their seed selection processes increased, especially when the cassava disease levels were under control.

Based on these findings, we increased the number of information packages available in both countries and distributed them according to the factors mentioned above.

10.6 Building the Model for Pest Management and Healthy Seed Management

As part of the Extension Team, we combined the above-mentioned activities and the lessons we learned from our experience with this project. We established a sustainable dissemination model for Cambodia and Vietnam (Fig. 10.11). This model

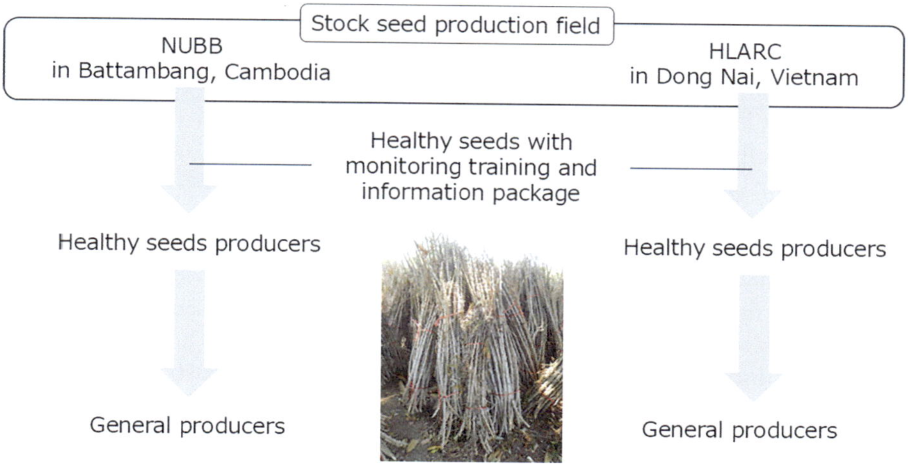

Fig. 10.11 Healthy cassava seed distribution model. (Illustration created by the authors)

is unique because it does not simply distribute healthy seeds to healthy seed producers but also enables them to produce healthy seeds by providing them with monitoring methods and information packages and allowing them to earn by selling their produce to general producers. With this model, trained experts at research and educational institutions have stock seed production fields to disseminate healthy seeds by monitoring training and information packages to healthy seed producers who can sell healthy cassava seeds to general producers.

10.7 Conclusion

Cassava is considered an important raw or processed cash commodity in Southeast Asian countries because of its starch and bioethanol potential. However, cassava is highly susceptible to viral diseases, such as witch broom and CMD, and harmful pests, such as mealybugs, red mites, and scale insects, owing to its genetic uniformity through propagation and relatively weak natural defense mechanisms against pests and pathogens. Thus, strengthening pest and disease control management and establishing a mechanism to provide healthy seeds are essential for minimizing potential damage to cassava.

To that end, as a part of a project, we aimed to combine the knowledge of the Plant Pathology, Pest Management, Seed Management, and Extension Teams to build a model for pest management and healthy seed management and share this knowledge with our research collaborators in Cambodia, Vietnam, and Thailand, as well as other countries facing the same conditions. The Plant Pathology Team studied cassava diseases and optimized the detection technology and monitoring systems. The Pest Management Team developed sustainable insect pest management systems and a simple tool for monitoring and biological control of cassava mealybugs using their natural enemies. The Seed Management Team identified useful varieties for breeding, developed new breeding technologies, and established a cultivation system for producing healthy cassava seeds. The Extension Team surveyed the current socioeconomic situation of cassava producers, which is widely recognized as a potentially significant factor affecting the adoption of newly introduced agricultural technologies and know-how, and developed a sustainable model for pest management and the extension of healthy cassava seeds.

As an Extension Team, we combined the above activities and the lessons learned from this project to establish a sustainable dissemination model for Cambodia and Vietnam (Fig. 10.11). We shared accumulated knowledge and trained the human resources of higher educational institutions and government agencies. Through the training of producers by disseminating information packages on disease and pest control and prevention, we believe that by empowering these human resources, the knowledge and know-how of disease control and prevention management can stay longer for the local experts of the educational institutions and government, as well as the cassava producer. Based on this model, they can react to new situations even when new pests or diseases occur in the future.

The project, which will be completed in 2022, established a model for healthy seed production and dissemination in Cambodia and Vietnam. However, this situation was not conducive to expanding the model as the CMD infection spread in Southeast Asia. To overcome this, trainer-to-trainer training could be beneficial to the governments of Laos and other Southeast Asian countries.

Development of resistant varieties is essential for pest and disease control and prevention. Research institutes in Vietnam and Thailand are vigorously developing CMD-resistant varieties. Once CMD-resistant varieties with high yields and starch contents are developed, the technologies and production and dissemination systems developed by the project can be used to increase the production and dissemination of these varieties.

During the implementation of the project, the government of Cambodia initiated the promotion of cassava processing industries under the "National Policy on Cassava 2020–2025" (Royal Government of Cambodia 2020). This declares cassava to be one of the potentially essential crops for agroindustrial development in Cambodia. The Ministry and its allied universities should promote and encourage propagation farms and private cassava stem producers to sell and distribute healthy, disease-free cassava stems. A member from NUBB also joined the official meeting and expressed an opinion based on the knowledge gained from the project. For the government of Vietnam, Decision No. 1748/QD-TTg dated December 30, 2023, of the Prime Minister approving the Strategy for Crop Development to 2030, with a vision to 2050 (Vietnamese Government 2023); Decision No. 1115/QD-BNN-TT dated April 17, 2024, of the Ministry of Agriculture and Rural Development approved the Project "Sustainable development of the cassava industry to 2030, with a vision to 2050" (MARD 2024). The Vietnamese government aims to strengthen its domestic processing plants and develop a stronger cassava industry.

Thus, cassava remains an important commodity for both Cambodia and Vietnam and likely for other Southeast Asian countries such as Laos, as does pest and disease control and management. In this regard, the model created by the project is cutting-edge to the latest trends in the cassava sector in Southeast Asia. We hope that the model developed in this project will be utilized in the future by introducing new resistant varieties and can contribute to local production and the economies of the countries concerned.

References

Alvarez E, Pardo JM, Mejia JF, Bertaccini A, Thanh ND, Hoat TX (2013) Detection and identification of 'Candidatus Phytoplasma asteris'-related phytoplasmas associated with a witches' broom disease of cassava in Vietnam. Phytopathogenic Mollicutes 3:77–81. https://hdl.handle.net/10568/51409

Baba T, Nomura H, Srean P, Than T, Ito K (2021) The possibility of producing and distributing healthy cassava seeds in Cambodia. Int J Environ Rural Dev 12(1):209–214. https://doi.org/10.32115/ijerd.12.2_209

Baba T, Nomura H, Srean P, Than T, Ito K (2022) Effects of mechanization and investments on the technical efficiency of cassava farms in Cambodia. Agriculture 12(4). https://doi.org/10.3390/agriculture12040441

Baba T, Nomura H, Than T, Srean P, Ito K (2023) How do small-scale cassava producers overcome global issues? Cassava profit and technical efficiency in Cambodia. J Agric Sci 15(8):1–15. https://doi.org/10.5539/jas.v15n8p1

Delaquis E, Andersen KF, Minato N, Le Cu TT, Karssenberg ME, Sok S, Wyckhuys KAG, Newby JC, Burra DD, Srean P, Phirun I, Le ND, Pham NT, Garrett KA, Almekinders CJM, Struik PC, de Haan S (2018) Raising the stakes: Cassava seed networks at multiple scales in Cambodia and Vietnam. Front Sustain Food Syst 2:73. https://doi.org/10.3389/fsufs.2018.00073

FAO (2024) Food and agriculture statistics, Retrieved from https://www.fao.org/faostat/en/#home. Accessed on 15 Oct 2024

Ito K, Nakatani A, Baba T, Nomura H, Than T, Srean P (2021) Status of cassava production and distribution channels in Cambodia: prospects for sustainable production. Int J Environ Rural Dev 12(2):41–48. https://doi.org/10.32115/ijerd.12.2_41

MARD (2024) The Ministry of Agriculture and Rural Development of Vietnam Decision No. 1115/QĐ-BNN-TT: Sustainable development of the cassava industry to 2030, with a vision to 2050. https://vukehoach.mard.gov.vn/DataStore/News/20244225-Q%C4%90%20%C4%90%E1%BB%81%20%C3%A1n%20s%E1%BA%AFn.pdf

Moun S, Nomura H, Baba T, Than T, Ito K, Srean P (2021) Experimental study of extension impact on farmers' KAP towards Sri Lankan cassava mosaic disease prevention in Cambodia. Int J Environ Rural Dev 12(1):79–87. https://doi.org/10.32115/ijerd.12.1_79

Moun S, Kaewrahun S, Janket A, Nomura H, Baba T, Ito K, Jutagate T (2024) Experimental study of extension intervention on farmers' perception of willingness to pay (WTP) for healthy seeds among Cambodian cassava farmers. Int J Agric Technol 20(1):177–196

Nguyen NC, Baba T, Nomura H, Ito K (2021) The current status of cassava producers and healthy cassava seed production and distribution trial in Vietnam. Int J Environ Rural Dev 12(1):215–222. https://doi.org/10.32115/ijerd.12.2_215

Nguyen NC, Baba T, Nomura H, Ito K (2023) Household income diversity of small-scale cassava producers in Vietnam. Int J Environ Rural Dev 14(1):139–146. https://doi.org/10.32115/ijerd.14.1_139

Royal Government of Cambodia (2020) Policy on Cassava 2020–2025. Royal Government of Cambodia, Phnum Penh, Cambodia

Tokunaga H, Baba T, Ishitani M, Ito K, Kim O-K, Ham LH, Le HK, Maejima K, Namba S, Natsuaki KT, Van Dong N, Nguyen HH, Nguyen NC, Nguyen AV, Nomura H, Seki M, Srean P, Tanaka H, Touch B, Trinh HX, Ugaki M, Uke A, Utsumi Y, Wongtiem P, Takasu K (2018) Sustainable management of invasive pests of cassava in Vietnam, Cambodia and Thailand. In: Crop production under stressful conditions: application of cutting-edge science and technology in developing countries. Springer, Singapore, pp 131–157

Uke A, Hoat TX, Quan MV, Liem NV, Ugaki M, Natsuaki KT (2018) First report of Sri Lankan cassava mosaic virus infecting cassava in Vietnam. Plant Dis 102(12). https://doi.org/10.1094/PDIS-05-18-0805-PDN

Vietnamese government. 2023 Decision No. 1748/QD-TTg: Strategy for Crop Development to 2030, with a vision to 2050. https://datafiles.chinhphu.vn/cpp/files/vbpq/2024/01/1748-ttg.signed.pdf

Wang HL, Wang XW, Cui XY, Liu SS (2015) First report of Sri Lankan cassava mosaic virus infecting cassava in Cambodia. Plant Dis 100(5):1029. https://doi.org/10.1094/PDIS-10-15-1228-PDN

Impact and Control of Transboundary and Invasive Banana (Musa spp.) Wilt Pathogen, Fusarium oxysporum f. sp. cubense

11

Nobumitsu Sasaki, Satoshi Yokoi, Carla Ximena Trigoso Hidalgo, Sakae Suzuki, Sakura Takahashi, Kei-ai Sha, Toshiyuki Fukuhara, Yoshihiro Nomura, Masaaki Yamada, Haruka Watanabe, Takeshi Kashiwa, Motoichiro Kodama, Kazutaka Kido, Tomoko Abe, Andrés Virgilio Casas-Díaz, Lilliana María Aragón-Caballero, Angel Alfonso Palomo-Herrera, María del Carmen Gonzáles-Miranda, Yuichi Nishida, Cecilia Inés Ono-Morikawa, Dina Lida Gutiérrez-Reynoso, Rosa María Cabrera-Pintado, Esmeralda Flor Cáceres-Patricio, Oscar Esmael Cabezas-Huayllas, Victorino Rivas-Pulache, Julio Alfonso Chia-Wong, and Tsutomu Arie

N. Sasaki · S. Yokoi · C. X. T. Hidalgo · S. Suzuki · S. Takahashi · K.-a. Sha · T. Fukuhara · Y. Nomura · M. Yamada · H. Watanabe · T. Arie (✉)
Tokyo University of Agriculture and Technology (TUAT), Fuchu, Tokyo, Japan
e-mail: chaki@cc.tuat.ac.jp; s235805z@st.go.tuat.ac.jp; s247700w@st.go.tuat.ac.jp; ssakae@cc.tuat.ac.jp; tsakura@go.tuat.ac.jp; fuku@cc.tuat.ac.jp; ny318@cc.tuat.ac.jp; masaakiy@cc.tuat.ac.jp; fw3397@go.tuat.ac.jp; arie@cc.tuat.ac.jp

T. Kashiwa
Japan International Research Center for Agricultural Sciences (JIRCAS), Tsukuba, Ibaraki, Japan
e-mail: kashiwat0417@jircas.go.jp

M. Kodama · K. Kido
Tottori University, Tottori, Japan
e-mail: mk@tottori-u.ac.jp; kazu-kido@tottori-u.ac.jp

T. Abe
RIKEN, Wako, Saitama, Japan
e-mail: tomoabe@riken.jp

A. V. Casas-Díaz · L. M. Aragón-Caballero · A. A. Palomo-Herrera · M. del Carmen Gonzáles-Miranda · C. I. Ono-Morikawa
Universidad Nacional Agraria La Molina (UNALM), La Molina, Lima, Peru
e-mail: cda@lamolina.edu.pe; lili@lamolina.edu.pe; fonz@lamolina.edu.pe; mcgonzales@lamolina.edu.pe; ceciliaono@lamolina.edu.pe

Y. Nishida
Universidad Nacional Agraria La Molina (UNALM), La Molina, Lima, Peru

Japan International Cooperation Agency (JICA), Chiyoda, Tokyo, Japan
e-mail: ynishida@lamolina.edu.pe

T. Nagamine et al. (eds.), *Plant Genetic Resources (PGR) for Sustainable Crop Production*, Sustainable Development Goals Series, https://doi.org/10.1007/978-981-96-7117-5_11

Abstract

Fusarium wilt (FW), or Panama disease, which is caused by *Fusarium oxysporum* f. sp. *cubense* (*Focb*), is one of the most destructive threats to banana/plantain (*Musa* spp.) production worldwide. Especially the pandemic of *Focb* race TR4, which can cause FW on the FW-resistant cultivar 'Cavendish', is damaging banana plantations in Asia, India, Australia, East Africa, and South America.

In the Selva area of Peru, banana cultivation is an important part of farmers' livelihoods. A team of Japanese and Peruvian research institutes is currently working on a SATREPS project with the goal of not only preventing the introduction of *Focb* race TR4 into Peruvian organic banana production areas but also enabling the continuation of high-quality banana production by preventing existing FW, thereby improving the livelihoods of farmers in the region. To achieve this goal, five research activities are conducted in the jungle, Tingo María area: (1) establishment of specific diagnosis technology for FW at the macro- to micro-level; (2) development of varieties that are resistant to FW using mutation induction; (3) establishment of a pathogen-free seedling production and supply system; (4) analysis of the microbial flora of disease-suppressing soil; and (5) establishment of low-environmental-impact FW control technology, such as biological pesticides and plant activators.

Keywords

Banana · Wilt disease · Fusarium oxysporum · Pandemic · Detection · Resistance · Pathogen-free seedlings · Microbiome · Biocontrol

11.1 Introduction

Bananas and/or plantains are plant species in the genus *Musa* L. that originated from Southeast Asia, including Indomalaya, and are now cultivated in over 130 countries in the so-called "Banana belt" between the 30 degrees latitudes. The world's banana fruit production was 13.9 million tons in 2023 (FAOSTAT), including dessert bananas, or fruit bananas, and cooking bananas, or plantains.

One of the most destructive threats to banana production is the fusarium wilt (FW), or Panama disease, which is caused by *Fusarium oxysporum* Schlecht. emend. Snyd. et Hans. f. sp. *cubense* Snyd. et. Hans (*Focb*) (Hwang and Ko 2007; Ploetz and Correll 1988; Snyder and Hansen 1940). The pathogen is classified into three races: 1, 2, and 4, based on its virulence in specific banana cultivars (Ploetz 1990, 1994, 2006). In the 1950s, the exports of *M. acuminata* Colla cv. "Gros Michel' (AAA triploid genome type) from Central America to North America and Europe dramatically decreased due to an epidemic of *Focb* race 1. Race 1 damages not only cv. 'Gros Michel' but also cvs. 'Silk' (AAB), 'Pome' (AAB) and etc. Race 2 causes endemic disease in several cooking banana cultivars such as 'Bluggoe' (ABB) in Central America (Su et al. 1986). Due to the significant impact of *Focb* race 1 on cv. 'Gros Michel', this variety has been replaced with cv. 'Cavendish' (AAA), which is resistant to races 1 and 2 (Nitani et al. 2018; Su et al. 1986; Dita et al. 2010). Currently, cv. 'Cavendish' holds about 99% of the banana trade (Perez-Vicente et al. 2014). In the 1960s, race 4 capable to infect cv. 'Cavendish' together with cultivars susceptible to races 1 and 2 was reported in subtropical areas such as Taiwan (Su et al. 1986; Zhang et al. 2013), the Philippines and the Canary Islands, and is now designated as

D. L. Gutiérrez-Reynoso · R. M. Cabrera-Pintado · E. F. Cáceres-Patricio
Instituto Nacional de Innovación Agraria (INIA), La Molina, Lima, Peru
e-mail: dgutierrez@inia.gob.pe; rcabrera@inia.gob.pe

O. E. Cabezas-Huayllas · V. Rivas-Pulache · J. A. Chia-Wong
Universidad Nacional Agraria de la Selva (UNAS), Tingo María, Peru
e-mail: oscar.cabezas@unas.edu.pe; victorino.rivas@unas.edu.pe

subtropical race 4 (race SR4) (Hwang and Ko 2007). Since the late 1980s, devastating damage from FW in cv. 'Cavendish' has been reported in tropical areas, such as Indonesia and Malaysia. The causal *Focb* of this severe damage was genetically distinct from race SR4 and was designated as tropical race 4 (race TR4) (Smith et al. 2014). TR4, at present, is the most devastating race of *Focb*, and its pandemic has damaged banana plantations in Asia, including the Philippines, India, Australia, East Africa, and South America, such as Colombia in 2019, Peru in 2021, and Venezuela in 2023 (Acuña et al. 2022; García Bastidas et al. 2020; Munhoz et al. 2024).

In Peru, bananas are mainly cultivated in the northern, coastal, and jungle regions. Large plantations cultivating "Cavendish" are settled in Tumbes and Piura areas, in the northern region. In the coastal and jungle regions, a lot of small farmers produce various cultivars such as "Isla', 'Manzano', 'Bellaco', 'Guayabo', 'Morado', 'Seda', and 'Moquicho', which include dessert cultivars (banana) and cooking cultivars (plantain). Although FW caused by race TR4 is restricted in the northern region (Munhoz et al. 2024), damage to bananas by FW races 1 and SR4 has been existed in all banana-growing regions (Fig. 11.1). Marin (2024) conducted a study on 189 banana farms distributed in five production regions in Peru and found that there was a prevalence of 28.04% of FW in bananas, being higher in the Junín region, with 55.56%. He also reported that the cv. 'Isla' was the most susceptible. Banana cultivation can be increasingly part of a sustainable replacement of the coca (ca. *Erythroxylum coca* Lam. var. *ipadu* Plowman; a source of cocaine) cultivation in the jungle region. However, smallholder farmers are suffering because FW has reduced the production of bananas, an important source of income in the region.

Fig. 11.1 Banana wilt symptoms observed in a field in Tingo María area, Peru. (**A-C**), banana plants showing yellowing of the leaves and wilt, (**D**, **E**), cross section of a pseudostem from an infected plant. Browning of the vascular bundles and longitudinal pseudostem-spritting are evident

Then, researchers from Tokyo University of Agriculture and Technology (TUAT), Tottori University (Tottori U), and The Japan International Research Center for Agricultural Sciences (JIRCAS) in Japan have teamed up with researchers from Universidad Nacional Agraria La Molina (UNALM), Instituto Nacional de Innovación Agraria (INIA), and Universidad Nacional Agraria de la Selva (UNAS) in Peru and are conducting international joint collaboration research with the goals of controlling banana FW and improving the quality of bananas in the jungle region, as well as improving the living standards of small-scale banana growers in the jungle region. This study is conducted as part of the Science and Technology Research Partnership for Sustainable Development (SATREPS) project titled "The Project on Establishment of an Alert

System for-*Fusarium oxysporum* f. sp. *cubense*, the Banana and Plantain Wilt Pathogen, and Mitigation Strategy of the Disease." The project is carried out with the support of the Japan Science and Technology Agency (JST) and Japan International Cooperation Agency (JICA).

To achieve this goal, five research activities are being conducted in the jungle, Tingo María area: (1) establishment of specific diagnosis technology for FW at the macro- to micro-levels; (2) development of varieties that are resistant to FW using mutation induction; (3) establishment of a pathogen-free seedling production and supply system; (4) analysis of the microbial flora of disease-suppressing soil; and (5) establishment of low-environmental-impact FW control technology, such as biological pesticides and plant activators. In this chapter, the results obtained from research activity 1 are mainly described.

11.2 Research Activity 1: Diagnosis and Alert System for FW Occurrence at the Macro- and Micro-Levels

Banana fields in the Tingo María area are distributed from Huallaga riverbanks to mountain slopes, making it difficult to regularly approach banana individual plants and conduct health examinations. Therefore, we aimed to first grasp individuals or populations suspected of FW from a macro perspective by image analysis using multispectral images collected from drones (Fig. 11.2) and then apply micro-level technology to precisely diagnose the pathogen of each FW-suspected individual (Fig. 11.2).

To date, we have identified genes characteristic of each *Focb* race based on whole genome data (Asai et al. 2019). In comparison of the genome data between *Focb* isolates, we clarified *Focb*- and/or its race-specific genes and designed Loop-mediated Isothermal Amplification (LAMP) primer sets for specific determination of *Focb* and its races, which are now sold by Nippon Gene Co. (https://www.nippongene.com/kensa/products/lamp-primer/banana/foc.html; Table 11.2 and Fig. 11.3).

From 2022 to 2024, we visited fields in the Tingo María area and sampled cvs. 'Bellaco', 'Guayabo', 'Isla', 'Manzano', 'Moquicho', and 'Seda' plants presenting typical FW symptoms, such as yellows, wilt, vascular-browning, and pseudostem-spritting were recorded (Table 11.1 and Fig. 11.1), and the vascular tissues with browning were sampled to bring to the laboratories in UNALM and UNAS.

A Fusarium-selective medium, Fo-G1 (Nishimura 2007), was used to isolate the FW pathogen from sampled banana tissues. A small piece of the vascular tissue with browning symptoms was placed on a Fo-G1 plate and incubated for several days at 27 °C. A small colony-edge fragment was transplanted onto fresh Fo-G1 medium. After repeated transplantation, a pure culture was established, maintained on potato dextrose agar (PDA), and used for further analyses.

For the pathogenicity test, each isolate was cultured in potato dextrose broth (PDB) medium at 27 °C by shaking (120 spm) for about 3 days. The obtained bud-cells (yeast like spore cells) were collected by centrifugation and suspended in distilled water to ca. 10^8 spores/ml. Approximately 15 cm-tall banana plants of cvs. 'Shima-banana' (having no resistance; Ryukashoji, Uruma, Okinawa, Japan) and 'Cavendish' (having resistance to races 1 and 2; Dole Philippines, South Cotabato, Philippines) were planted on sterilized soil and were inoculated with each isolate by pouring the bud-cell suspension into the soil, kept in a greenhouse at about 25 °C for ca. 40 days, and the symptoms (leaf-yellowing and vessel-browning) of each plant were observed and evaluated.

DNA was purified from each fungal culture using ISOIL soil DNA purification kit (Nippon Gene). A small piece of mycelia formed on the plate medium was taken and placed into the 100 μl extraction solution (ES) in a PCR tube, heated at 95 °C for 10 min, and put on ice for 2 min. Instead of mycelia, banana tissues and field soil can be used for DNA purification.

LAMP was performed using purified DNA as a template, DryADD LAMP Master Mix (Turbidity/Visible Dye) (Nippon Gene), and the primers. In a tube containing DryADD LAMP

Fig. 11.2 An example of a drone image of a banana field (cv. Apple; photo taken by Dr. J. Yoneda in Kin, Okinawa, Japan). The suspected fusarium wilt (FW) individuals are indicated with white arrowheads

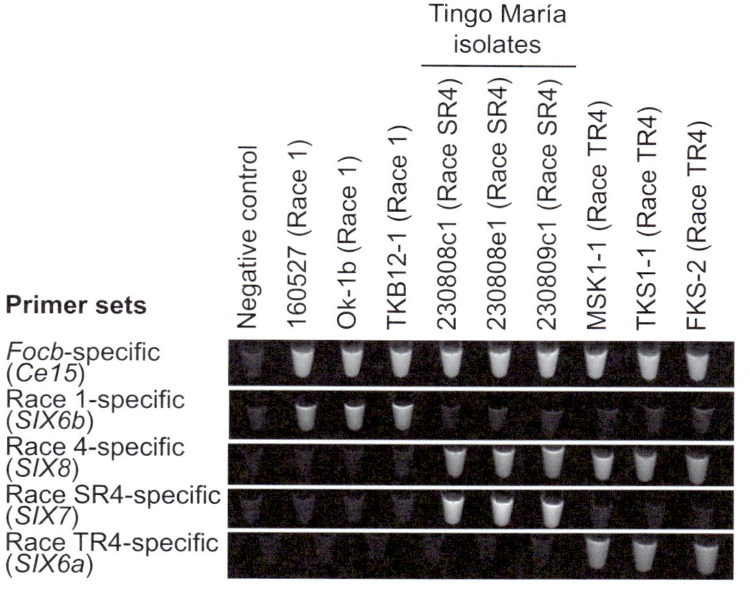

Fig. 11.3 Example of the LAMP reaction. Tubes that appear white are positive, and tubes that do not appear white are negative. The reaction took only 30–50 min. Three isolates from Tingo María area are determined to be *Focb* race SR4

Master Mix, 15.5 µl ddWater, 5.0 µl LTV Dissolve Solution, 2.5 µl 10 × LAMP Primer Mix, and 2 µl template DNA were mixed and incubated at 64.5 °C. LAMP, within 30–50 min. We revealed that *Focb* occurring in the Tingo María area were dominantly of race SR4, and a small number were of race 1 (Tables 11.1, 11.2, and Fig. 11.3) using this LAMP. We also confirmed that race TR4 has not yet invaded the Tingo María area.

We are also working to disseminate this technology by conducting training courses at UNALM, UNAS, and other collaborative research partners in Peru (Fig. 11.4).

Table 11.1 Examples of banana field and samples in Tingo Maria area and the pathogen and race diagnosis

Sample						Pathogenicity on		Determination of race by	
Field no.	Plant no.	Cultivar	FW Symptom	Isolation of *F. oxysporum*	Isolate no.	cv. Shima-banana	cv. Cavendish	PCR	LAMP
Samples from Ting Maria Region									
TM02	220909b[a]	Guayabo	Y[b]	Y[c]	1	nt[d]	nt[d]	1[e]	1[e]
					2	nt	nt	SR4	SR4
					3	nt	nt	SR4	SR4
	220909c	Seda Punta	Y	Y	2	nt	nt	SR4	SR4
	241114a	Seda Punta	N	N	nt	nt	nt	nt	N
	241114b*	Seda Punta	Y	Y	4	nt	nt	nt	SR4
TM03	220910a	Isla	Y	Y	1	nt	nt	SR4	SR4
TM04	220910c	Manzano	Y	Y	1	nt	nt	SR4	SR4
					2	nt	nt	1	1
TM05	220910d	Guayabo	Y	Y	1	nt	W	SR4	SR4
TM06	230808a*	Isla Maleno	N	N	–	nt	nt	nt	N
	230808b*	Isla Maleno	N	N	–	nt	nt	nt	N
	230808c	Isla Maleno	Y	Y	1	Y	W	SR4	SR4
TM07	230808e	Isla Maleno	Y	Y	1	nt	nt	nt	SR4
TM08	230809c	Manzano	Y	Y	1	Y	W	SR4	SR4
	230809d*	Manzano	N	N	–	nt	nt	nt	N
TM24	241118c*	Seda	N	N	–	nt	nt	nt	N
	241118d	Seda	Y	Y	1	nt	nt	nt	SR4
Control									
Race 1					160,527	Y	N	1	1
Race SR4					23TM22-4	Y	W	SR4	SR4
Race TR4					MSK1-1	Y	Y	TR4	TR4

[a]The six digits in the plant number indicate the date of sampling and an alphabet indicates each individual plant
[b]Y, the plant exhibit the symptoms of FW; N, the plant did not present symptoms of FW
[c]Y, *F. oxysporum* was isolated from the plant; N, no *F. oxysporum* was isolated from the plant
[d]Y, the plant presented symptoms; W, the plant presented weak symptoms; N, the plant presented no symptoms; nt, not tested
[e]By PCR or LAMP the pathogen was determined to be race 1 (1), race SR4 (SR4) or not *Focb* (N). nt, not tested
*indicates possible disease suppressive soil in the field

Table 11.2 Reaction of each *F. oxysporum* race with LAMP primer sets based on genes on an accessory chromosome

Primer set (gene on which the primer set was designed)	*F. oxysporum* f. sp. *cubense*			Other forms
	Race 1	Race 4		
		Race SR4	Race TR4	
Focb-specific (*Ce15*)	+	+	+	–
Race 1-specific (*SIX6b*)	+	–	–	–
Race 4-specific (*SIX8a*)	–	+	+	±
Race SR4-specific (*SIX7*)	–	+	–	±
Race TR4-specific (*SIX6a*)	–	–	+	–

+, positive; –, negative, ±, positive or negative

Fig. 11.4 Example of educational activities. Demonstration determination of banana wilt pathogen and its races by LAMP. (**A**) For researchers/students from UNALM, INIA, and UNAS (in UNALM plant clinic laboratory). (**B**) For farmers/farmer's disciples (in a banana farm in Tingo María area)

11.3 Research Activity 2: Generation of New Banana Varieties Resistant to FW by Inducing Mutations Through Heavy Ion Beam and Gamma Irradiation

Edible bananas are triploid and do not form seeds; thus, the breeding of FW- resistant/tolerant cultivars through crossbreeding is impossible. Therefore, this research activity aims to obtain resistant/tolerant plants to FW by accession selection or mutagenesis.

In the first study, disease resistant/tolerant banana accessions are going to be selected from the field collection of the germplasm bank maintained by INIA. Fifty-one accessions are tissue cultured and stably in vitro propagated. The FW-resistance/tolerance of each accession will be evaluated in further experiments.

In the second study, tissue-cultured shoots are mutagenized by radioactive irradiation (Fig. 11.5). Four cultivars, 'Cavendish' (Subgroup Cavendish, genetic group AAA), 'Bellaco Harton' (Subgroup Plantain, genetic group AAB), 'Bellaco Plantano' (Subgroup Plantain, genetic group AAB), and 'Isla' (Subgroup Iholena, genetic group AAB) which were obtained from Tingo María grower's fields, in vitro established and multiplied and then subjected to mutation. We generated mericlonal plantlets of bananas and plantains, and the in vitro 'Isla' shoots were exposed to a gamma ray (Gamma Co-60, Gamma beam 127 IR 194 type II Panoramic Irradiator; Servicio Nacional de Sanidad Agraria del Perú (SENASA), La Molina, Peru) or in vitro 'Cavendish', 'Isla', 'Bellaco Harton', and 'Bellaco Plantano' shoots were exposed to heavy ion beam (Carbon; RIKEN Nishina Center, Wako, Japan). Based on the median lethal inten-

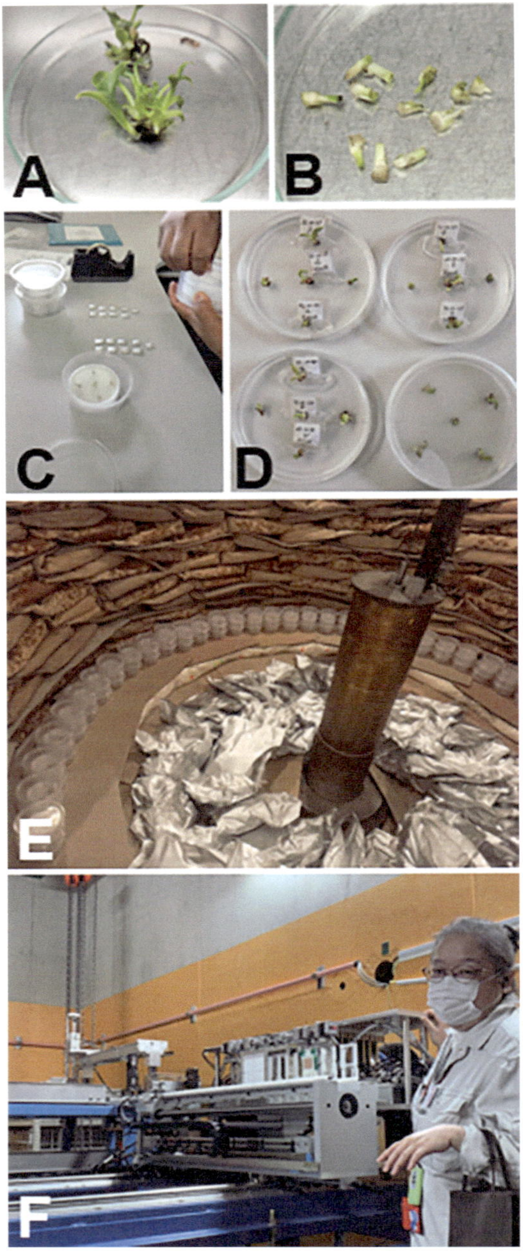

Fig. 11.5 Banana plantlets were exposed with gamma ray and heavy ion beam for mutation. (**A–D**) Banana plantlets used in the study. The small shoots were put on medium in Petri dishes. (**E**) The shoots were exposed with gamma ray (Gamma Co-60, Gamma beam 127 IR 194 type II Panoramic Irradiator; SENASA, La Molina, Peru). (**F**) The plantlets were exposed with heavy ion beam (Carbon; RIKEN, Wako, Japan)

sity, the optimal irradiation intensities of the gamma ray and heavy-ion beams were set to 40 and 30 Gy, respectively. The irradiated buds were then grown in vitro to the third generation through the chimera dissolution process. After propagation and acclimatization of the plantlets, FW-resistant/tolerant lines will be screened using an inoculation test with *Focb* race SR4 (INIA) and race TR4 (TUAT).

11.4 Research Activity 3: Publicizing the Importance of Using Pathogen-Free Seedlings and Establishing a Supply System for Them

Focb can be transmitted by seedling, irrigation water, flood, agricultural equipment, and human feet. *Focb* race TR4 is causing a global pandemic, in molecularly almost single lineage, indicating that *Focb* race TR4 is a single origin and has spread worldwide via mainly seedlings. Farmers can plant seedlings from the base of banana parent plants grown in the field at no cost, but there is concern that latent *Focb* may be transmitted by the seedlings. Moreover, we diagnosed two seedlings generated from symptom-free axillary shoots obtained in the fields in Tingo María area by LAMP, as examples, and found one of them infested with *Focb* race 1 (Table 11.3). Therefore, we propose the farmers to understand the importance of using pathogen-free seedlings generated from mericlones, and Peruvian institutions are going to work together to supply these pathogen-free seedlings to farmers. The LAMP test developed in the research activity 1 (Fig. 11.3) will also be used to guarantee that the seedlings are *Focb*-free. A demonstration plot with plants of the cultivar 'Isla' from in vitro culture will be installed in Tingo María area, to demonstrate to farmers the benefits of using pathogen-free seedlings.

Table 11.3 *Foc* detection from seedlings from symptom-free axillary shoots

	Isolation on medium		Detection by LAMP using the Primer Sets				
			Foc	*Foc* race 4	*Foc* race SR4	*Foc* race TR4	
Plant sample	PDA	Nishimura[a]	*Ce15*	*SIX8a*	*SIX7*	*SIX6a*	Diagnosis
A01	–	–	–	–	–	–	Healthy
C02	+	+	+	–	–	–	*Focb* race 1

[a]A *Fusarium*-selective medium

11.5 Research Activity 4: Analysis of Microbial Ecosystems in Disease-Suppressive Soils (DSSs)–Identification of Effective Microbiota (Microbial Ecosystem) That Constitutes DSSs

Soil that is unlikely to cause disease despite the presence of pathogens is called "disease-suppressive soil (DSS)." The overall goal of this research activity is the characterization of DSSs to FW in the Tingo María area. We collected soil samples from fields in the Tingo María area, where FW has not occurred (Table 11.1, indicated with *). We are currently testing the FW-suppressiveness of each soil by planting bananas and clarifying the microbial flora thought to be involved in disease suppression through metagenomic analyses. By the end of the project, it is expected that the physical, chemical, and biological properties of DSSs will be elucidated and that the results will be used to establish biocontrol agents for FW of bananas.

As model experiments, we identified microorganisms with inhibitory activity against *Focb* in banana fields in Okinawa, Japan, and found several bacterial strains with strong growth inhibitory activity against *Focb*, for instance, *Bacillus* spp. The results have been reported at the Annual Meetings of the Phytopathological Society of Japan (PSJ; Tomita et al. 2024). It is expected that these data will help to efficiently select and identify candidate microbial isolates in field samples in Peru.

11.6 Research Activity 5: Establishment of Environmentally Friendly FW Control Technology, Such as Biopesticides and Plant Activators

There is a global demand for environmentally friendly plant disease control technology. Banana cultivation in the Tingo María area is generally organic. Therefore, it is necessary to control FW using technologies with low environmental impact. Currently, we are considering the application of nonpathogenic *Fusarium* spp. (Saito et al. 2021) and of plant activators, such as Validamycin A (VMA; Ishikawa et al. 2005), to induce resistance in bananas to control FW. In the future, we would like to propose new biopesticide using synthetic microbial communities (SynCom) approach, which artificially reconstructs the microbial flora involved in disease suppression based on the results of the research activity 4.

11.7 Discussion

Our goal is to develop and implement a low-environmental-impact "comprehensive FW control package," consisted with elaboration of manuals, setting up a molecular diagnostic laboratory in Tingo María area, promotion and distribution of pathogen-free banana nursery plantlets for the farmers, students/researchers exchange and train experts to operate the "package," which will prevent the spread and epidemic of FW of bananas in Peru's tropical rainforests (i.e., Tingo María area), and increase the productivity of bananas, particularly organic bananas. This may improve the standard of living of small-scale farmers, while also reducing the burden on the environment and farmers. This might connect to the development of good agricultural practices. We believe that this "package" and human resources can be deployed in banana-producing regions around the world.

Acknowledgments This study has been conducted mainly with the support of the JST/JICA SATREPS and Grants-in-Aid for Scientific Research (20KK0133, 23H00331). Soil and microorganisms were exported with permission from SENASA, Peru, and imported with special permission from the Minister of Agriculture, Forestry and Fisheries (MAFF) of Japan.

The details of each research activity will be described in separated papers.

In recent years, ***F. oxysporum*** has been shown to be a species complex, and new species have been proposed for banana wilt fungus, too. In this study, however, we are using ***F. oxysporum*** as a traditional species complex and ***Focb*** as banana wilt pathogen.

References

Acuña R, Rouard M, Leiva AM, Marques C, Olortegui JA, Ureta C, Cabrera-Pintado RM, Rojas JC, Lopez-Alvarez D, Cenci A, Cuellar WJ, Dita M (2022) First report of *Fusarium oxysporum* f. sp. *cubense* tropical race 4causing fusarium wilt on Cavendish bananas in Peru. Plant Dis 106:2268

Asai S, Ayukawa Y, Gan P, Masuda S, Komatsu K, Shirasu K, Arie T (2019) High-quality draft genome sequence of *Fusarium oxysporum* f. sp. *cubense* strain 160527, a causal agent of Panama disease. Microbiol Resour Announc 8:e00654-19. https://doi.org/10.1128/MRA.00654-19

Dita MA, Waalwijk C, Buddenhagen IW, Souza MT Jr, Kema GHJ (2010) A molecular diagnostic for tropical race 4 of the banana fusarium wilt pathogen. Plant Pathol 59:348–357

FAOSTAT (2023) Crops and livestock products. https://www.fao.org/faostat/en/#data/QCL

García-Bastidas FA, Quintero-Vargas JC, Ayala-Vasquez M, Schermer T, Seidl MF, Santos-Paiva M, Noguera AM, Aguilera-Galvez C, Wittenberg A, Hofstede R, Sørensen A, Kema GHJ (2020) First report of fusarium wilt tropical race 4 in Cavendish bananas caused by *Fusarium odoratissimum* in Colombia. Plant Dis 104:994–994

Hwang S-C, Ko W-H (2007) Cavendish banana cultivars resistant to fusarium wilt acquired through somaclonal variation in Taiwan. Plant Dis 88:579–683

Ishikawa R, Shirouzu K, Nakashita H, Lee H, Motoyama T, Yamaguchi I, Teraoka T, Arie T (2005) Foliar spray of validamycin A or validoxylamine A controls tomato Fusarium wilt. Phytopathology 95:1209–1216

Marin D (2024) Intensidad y distribucion espacial de la Marchitez por fusarium (*Fusarium oxysporum* f. sp. *cubense*) de platanos y bananos en el Perú. Thesis for the master degree in plant pathology. UNALM, Lima, Perú, p 142

Munhoz T, Vargas J, Teixeira L, Staver C, Dita M (2024) Fusarium Tropical Race 4 in Latin America and the Caribbean: status and global research advances towards disease management. Front Plant Sci 15:1397617

Nishimura N (2007) Selective media for *Fusarium oxysporum*. J Gen Plant Pathol 73:342–348

Nitani T, Akai K, Hasegawa R, Ayukawa Y, Garcia RR, Chitose A, Komatsu K, Kikuno H, Natsuaki K, Arie T (2018) Panama disease of banana occurred in Miyakojima Island, Okinawa, Japan. J Gen Plant Pathol 84:165–168

Perez-Vicente L, Dita MA, Parte EM (2014) Technical manual: prevention and diagnostic of fusarium wilt (Panama disease) of banana caused by *Fusarium oxysporum* f. sp. *cubense* tropical race 4 (TR4). Food and Agriculture Organization (FAO), Rome

Ploetz RC (1990) Population biology of *Fusarium oxysporum* f. sp. *cubense*. In: Ploetz RC (ed) Fusarium wilt of banana. APS Press, St. Paul, pp 63–76

Ploetz RC (1994) Panama disease: return of the first banana menace. Intl J Pest Mgmt 40:326–336

Ploetz RC (2006) Fusarium wilt of banana is caused by several pathogens referred to as *Fusarium oxysporum* f. sp. *cubense*. Phytopathology 96:653–656

Ploetz RC, Correll JC (1988) Vegetative compatibility among races of *Fusarium oxysporum* f. sp. *cubense*. Plant Dis 72:325–328

Saito H, Sasaki M, Nonaka Y, Tanaka J, Tokunaga T, Kato A, Tran TTT, Van-V L, Le MT, Kanematsu S, Suzuki T, Kurauchi K, Fujita N, Teraoka T, Komatsu K, Arie T (2021) Spray application of nonpathogenic Fusaria onto rice flowers controls Bakanae disease (caused by *Fusarium fujikuroi*) in the next plant generation. Appl Environ Microbiol 87:e01959–e01920

Smith SF, Czislowski E, Meldrum RA, Zander M, O'Neill W, Balali GR, Aitken EAB (2014) Sequence variation in the putative effector gene *SIX8* facilitates molecular differentiation of *Fusarium oxysporum* f. sp. *cubense*. Plant Pathol 63:1

Snyder WC, Hansen HN (1940) The species concept in Fusarium. Am J Bot 27:64–67

Su HJ, Hwang SC, Ko WH (1986) Fusarial wilt of Cavendish bananas in Taiwan. Plant Dis 70:814–818

Tomita Y, Kido K, Arie T, Kodama M (2024) The search for useful microorganisms for biocontrol of Panama disease. Jpn J Phytopathol 90:52

Zhang H, Mallik A, Zeng RS (2013) Control of Panama disease of banana by rotating and intercropping with Chinese chive (*Allium tuberosum* Rottler): role of plant volatiles. J Chem Ecol 39:243–252

Natural Rubber: Its Importance and Problems Tackled with Indonesia Through SATREPS Program

12

Minami Matsui, Masahiko Ota, Emiko Okubo-Kurihara, Stephany Angelia Tumewu, Masafumi Shimizu, Tri Rapani Febbiyanti, Yoshiharu Yamamoto, Afdholiatus Syafaah, Retno Lestari, Sanya Khaerunnisa, Yuko Makita, Kiwamu Kase, and Achmad Eka Satria

Abstract

Natural rubber, primarily derived from *Hevea brasiliensis*, is a critical resource for a sustainable society due to its unique properties and indispensable applications in industries such as tire manufacturing, medical supplies, and industrial components. As the global demand for natural rubber continues to rise, ensuring the stability of its supply is increasingly important. However, *H. brasiliensis* faces significant threats due to its reliance on clonal propagation and diseases such as leaf-fall disease, which jeopardize supply security.

To protect natural rubber as a genetic resource, long-term efforts are required to enhance resistance to leaf-fall disease, alongside breeding and conservation initiatives. Additionally, the conservation of rubber trees

M. Matsui (✉)
RIKEN Center for Sustainable Resource Science (CSRS), Yokohama, Kanagawa, Japan

Yokohama City University, Yokohama, Japan
e-mail: minami@riken.jp

M. Ota
Nagasaki University, Nagasaki, Japan
e-mail: masahikoota@nagasaki-u.ac.jp

E. Okubo-Kurihara
RIKEN Center for Sustainable Resource Science (CSRS), Yokohama, Kanagawa, Japan

Department of Biology, Keio University, Yokohama, Japan
e-mail: ekurihara@keio.ac.jp

S. A. Tumewu · M. Shimizu
Faculty of Applied Biological Sciences, Gifu University, Gifu, Japan
e-mail: stephany.angelia.e7@f.gifu-u.ac.jp; shimizu.masafumi.f3@f.gifu-u.ac.jp

T. R. Febbiyanti
Plant Protection Laboratory, Indonesian Rubber Research Institute, South Sumatera, Indonesia

Y. Yamamoto · A. Syafaah
Gifu University, Gifu, Japan
e-mail: yamamoto.yoshiharu.h5@f.gifu-u.ac.jp

R. Lestari · S. Khaerunnisa · A. E. Satria
Universitas Indonesia, Kota Depok, Jawa Barat, Indonesia
e-mail: sanya.khaerunnisa@alumni.ui.ac.id

Y. Makita
Faculty of Engineering, Maebashi Institute of Technology, Maebashi, Gunma, Japan
e-mail: makita@maebashi-it.ac.jp

K. Kase
RIKEN Center for Advanced Photonics, Wako, Saitama, Japan
e-mail: kiwamu@riken.jp

T. Nagamine et al. (eds.), *Plant Genetic Resources (PGR) for Sustainable Crop Production*, Sustainable Development Goals Series, https://doi.org/10.1007/978-981-96-7117-5_12

not only ensures sustainability but also strengthens local economies that depend on rubber cultivation.

This chapter discusses the current status of natural rubber supply and *H. brasiliensis* cultivation, the challenges posed by leaf-fall disease, and our approach to these issues through scientific and integrated methods. By conducting innovative research and taking proactive actions, we aim to secure the long-term sustainability of the natural rubber industry while addressing environmental, economic, and social challenges.

Keywords

Indonesia · Hevea brasiliensis · Rubber tree · Leaf-Fall disease · Chemical fungicide · Biofungicide · Genome-based breeding · Transcriptome · Drone imaging · Mathematical modeling · Short visiting course · Bioinformatics

12.1 Current Status and Issues of Natural Rubber Production

12.1.1 Natural Rubber Production by Smallholder Farmers in Indonesia

Natural rubber, produced from para rubber trees (*Hevea brasiliensis)*, is one of the most renowned and significant agricultural commodities globally. Although para rubber trees originate from Brazil, Southeast Asia has become the primary production region (FAO 2024).

Natural rubber can be produced from large-scale plantations managed by companies or from smallholder farmers, resulting in diverse production structures across producing countries. Indonesia is characterized by a high presence of smallholder rubber production. As of 2022, smallholders account for 92.3% of dry rubber production (Statistics Indonesia 2022). A high proportion of smallholder production underscores the significant role of natural rubber in local livelihoods.

A leaf fall disease (LFD) caused by a fungus, *Pestalotiopsis*, has spread across Southeast Asia since 2017. Indonesia is one of the most severely affected countries. In the following, results are presented of the survey conducted in September 2023 by SATREPS project members as a baseline study of the conditions of rubber farmers under the LFD. We also refer to the context in which Indonesian rubber smallholders are situated.

Rubber prices in Southeast Asian countries experienced a boom during 2008–2011. However, after 2012, rubber prices significantly decreased and have generally remained low. For smallholders, low international prices result in low local prices, from which they suffer.

Natural rubber production and exports in Indonesia have declined since 2018. This decrease in production can be attributed, at least in part, to the impact of the *Pestalotiopsis* LFD. Over the long term, persistent low rubber prices have led to many rubber plantations being left unmanaged or even abandoned (Syarifa et al. 2023). These situations have created a sense that the natural rubber sector is a sunset industry. Starting from 2022, the Indonesian Government ceased providing fertilizer subsidies for rubber.

Meanwhile, there has been an increasing adoption of collective selling methods through Rubber Processing and Marketing Units (RPMUs: in Indonesian *Unit Pengolahan dan Pemasaran Bokar*). RPMUs typically conduct auctions for buyers of raw rubber materials to secure fair selling prices. Through RPMUs, smallholders also have access to various forms of assistance.

In September 2023, a team of the SATREPS project conducted a household survey using a questionnaire in a face-to-face manner in South Sumatra Province. We selected 10 villages: five from Banyuasin Regency and five from Muara Enim Regency. In each village, around 10 rubber farmers were requested to answer interviews. In total, we secured 103 respondents across all villages. We also conducted explanatory sessions on the LFD in combination with the household survey; scientific knowledge regarding *Pestalotiopsis* sp. and its potential impact was presented to ordinary farmers.

Respondents reported that due to the *Pestalotiopsis* LFD, natural rubber latex production had decreased on average by 47.7%, with values ranging from a maximum of 75.0% to a minimum of 17.0%.

The average size of rubber gardens owned was 2.6 ha, accounting for 92.9% of total land owned. Other land uses included horticulture, bush, and oil palm gardens, though these were relatively insignificant compared to rubber cultivation.

The average annual gross income among the surveyed households was Rp. 43,156,816 in 2022. Income from rubber averaged Rp. 28,350,990, constituting 65.7% of the total income. The degree of dependence on rubber income varied among households. Forty-one households (39.8%) relied solely on rubber income. The livelihood impact of the *Pestalotiopsis* LFD is considered to have been more pronounced among these households.

When asked about measures taken to control the *Pestalotiopsis* LFD, 95 households (92%) reported taking no action. Eight households (8%) had implemented some measures, including applying fertilizer, fogging with fungicide, land cleaning, smoking, and seeking support. However, all respondents indicated that their efforts were ineffective against *Pestalotiopsis* sp. According to t-tests, there was statistical significance between the groups of "measures taken" and "measures not taken" regarding total annual income ($p < 0.01$). Smallholders with higher incomes were more likely to take action against the disease. No systematic measures against the LFD by RPMUs were observed, although RPMUs are crucial organizations for local livelihoods in the rubber sector.

It was confirmed that the *Pestalotiopsis* LFD had significantly heightened the vulnerability of rubber smallholders' livelihoods, primarily due to the decrease in latex production.

12.1.2 World Natural Rubber Production and Influence of LFD

Along with the development of transportation, demand of natural rubber is increasing every year (Guerra et al. 2021; Herath et al. 2012). To meet the demand of natural rubber, production of natural rubber in rubber producing counties is increasing (Fig. 12.1). Japan imports more than 66% of TSR from Indonesia (FAO 2023). As from the academic name of natural rubber tree (*Hevea brasiliensis*), this tree originates from the Amazon basin of Brazil. Natural rubber tree was introduced to Sri Lanka (called Ceylon at that time) and East Asian countries in 1876. Now more that 90% of world supply is supported by East Asian countries.

Natural rubber production was active in Brazil early in the nineteenth century. Several companies such as Ford, Belterra, Goodyear made rubber plantations in Brazil and surrounding countries (Lieberei 2007). But all of them closed their factories because of an LFD, or South American Leaf-Blight (SALB). It attacked their plantations and made a devastating effect on their rubber plantations (Lieberei 2007). In 2017, another LFD was found in South Sumatra in Indonesia (Febbiyanti and Fairuza 2019). After that, many rubber producing countries in East Asian reported the same or similar LFD in their plantations. In 2019, around 11% of Indonesian plantations were affected by LFD (Febbiyanti and Fairuzal 2019). Also, Thailand, Malaysia, Sri Lanka, and Philippines reported natural rubber LFD with similar symptoms. By rubber research institutes of each country, it was reported that these LFD are caused by fungi, *Pestalotiopsis* sp., *Neopestalotiopsis* sp., *Pseudopestalotiopsis* sp., and *Collettrichum*. They are collectively called circular leaf spot disease (September 4–6, 2023, Colombo Sri Lanka). In 2023, in total 649,000 ha of rubber plantations were affected by LFD and showed apparent reduction of natural rubber production since 2018 (Fig. 12.1).

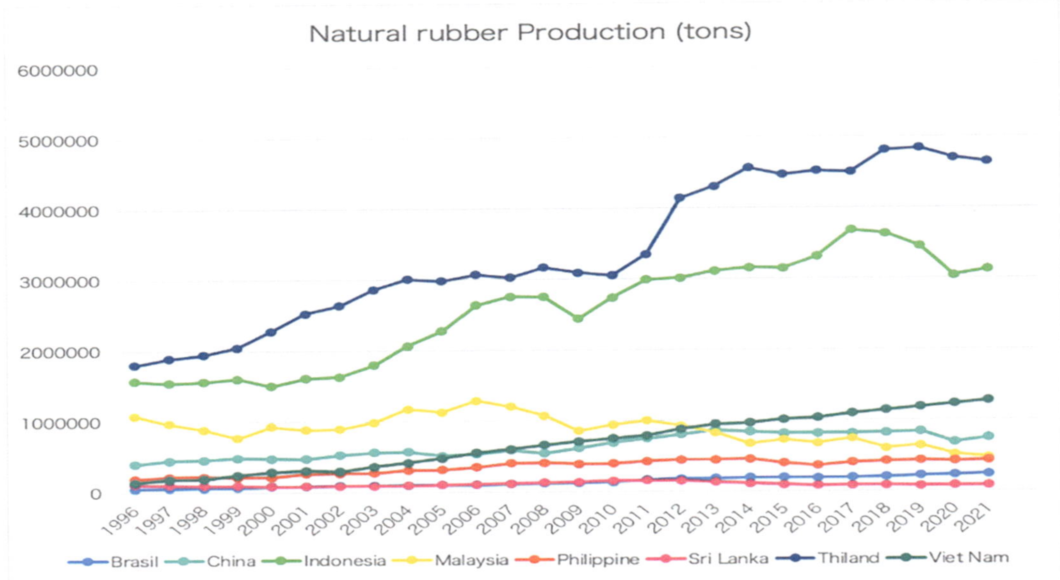

FAOSTAT data 20231123

Fig. 12.1 FAOSTAT data about annual production of natural rubber November 23, 2023

12.1.3 Current Status and Problems of Rubber Tree Leaf Fall Disease (LFD) in Rubber-Growing Countries of the World

Over the past four years, rubber production in Indonesia alone has decreased by 795,783 tons due to factors such as climate change, low prices, aging trees, and disease outbreaks (Premadasa 2020). Priyadarshan (2017) highlighted that leaf fall disease (LFD) has become a significant limiting factor for rubber plantations. Major fungal causes of LFD include *Corynespora cassiicola* (*Corynespora* LFD), *Oidium heveae* (powdery mildew), and *Colletotrichum gloeosporioides* (*Colletotrichum* LFD) (Mazlan et al. 2019). These pathogens occur in most rubber-producing countries and are widely studied globally (Chen et al. 2023). Furthermore, since 2017, a new form of LFD exhibiting different symptoms from the common LFD has been reported in rubber plantations in Indonesia. This new leaf fall disease is attributed to *Pestalotiopsis* and has now been widely reported in rubber plantations across Southeast Asia, India, China, and even Colombia (Darojat et al. 2023). This section outlines the main pathogens causing LFD in *Hevea brasiliensis*, particularly *Colletotrichum* and *Pestalotiopsis*, and reviews the characteristics of the disease caused by these fungi.

12.1.3.1 *Colletotrichum* LFD

The disease has been reported from southern India, Sri Lanka (Jayasinghe et al. 1997; Hunupolagama et al. 2017), Indonesia (Darojat et al. 2023), Malaysia, and Thailand (Pornsuriya et al. 2020). In India, the disease was first reported in 2017, followed by the spread of the disease to neighboring areas the following year in 2018; severe disease outbreaks were recorded during summer rains in May 2021 and May 2022. The heavily affected rubber plantations in Kerala recorded 80–95% defoliation and about 35% reduction in yield (Vineeth et al. 2024). From 2020 to 2023, leaf fall disease caused by *Colletotrichum,* particularly *Colletotrichum siamense* spread extensively in Thailand. This disease is characterized by the formation of spots and ulcerative lesions on the leaves and branches, leading to blackened and dead foliage. Infection with *Colletotrichum* spp. leads to secondary defoliation (SLF) in rubber trees. Young lesions produce conidiospores, while older lesions, when exposed to prolonged moisture, fail to form conidiospores. The extent of sporulation and fungal damage increases with prolonged leaf wetness and higher inoculum concentrations (Guyot et al. 2005) (Fig. 12.2).

12.1.3.2 *Neopestalotiopsis* Species LFD

Neopestalotiopsis species are significant fungi causing leaf spot disease in rubber trees, particularly reported in rubber-growing regions of Southeast Asia, including Thailand, Malaysia, and Indonesia (Damiri et al. 2022). This disease is marked by the appearance of small blackish-brown spots on leaves, eventually resulting in leaf drop. It tends to proliferate in warm, humid conditions. These pathogens primarily affect mature rubber leaves, causing symptoms such as necrotic spots, leaf yellowing, and leaf drop (Febbiyanti and Fairuza 2019). Recent research revealed that this new LFD is associated with *Pestalotiopsis* sp. (Febbiyanti and Fairuza 2019). Several species of *Pestalotiopsis* are related to the etiology of this new LFD, including *Neopestalotiopsis cubana* and *N. formicarum* (Pornsuriya et al. 2020), *N. aotearoa* (Li et al. 2021), and *P. microspora* (Kusdiana et al. 2020). In some cases, the new LFD is associated with a combination of pathogens, such as *Colletotrichum* sp. and *Pestalotiopsis* sp. (Aliya et al.2022), or other genera like *Calonectria* (Thaochan et al. 2020). Infection with the disease reduces the leaf area index and impacts latex rubber production (Kusdiana and Saputra 2022) (Fig. 12.3).

12.1.4 Conclusion and Future Perspectives

The spread of leaf fall disease in rubber trees has led to a reduction in rubber production by approximately 30–40%, highlighting the significant impact this disease has on global rubber production. The main causative pathogens, including *Corynespora*, *Colletotrichum*, *Oidium*, and *Pestalotiopsis*, exhibit varying levels of pathogenicity depending on environmental conditions, particularly humidity and temperature. Three key issues have emerged. First, it has been observed that the severity of infection increases primarily during the rainy season and with rising tempera-

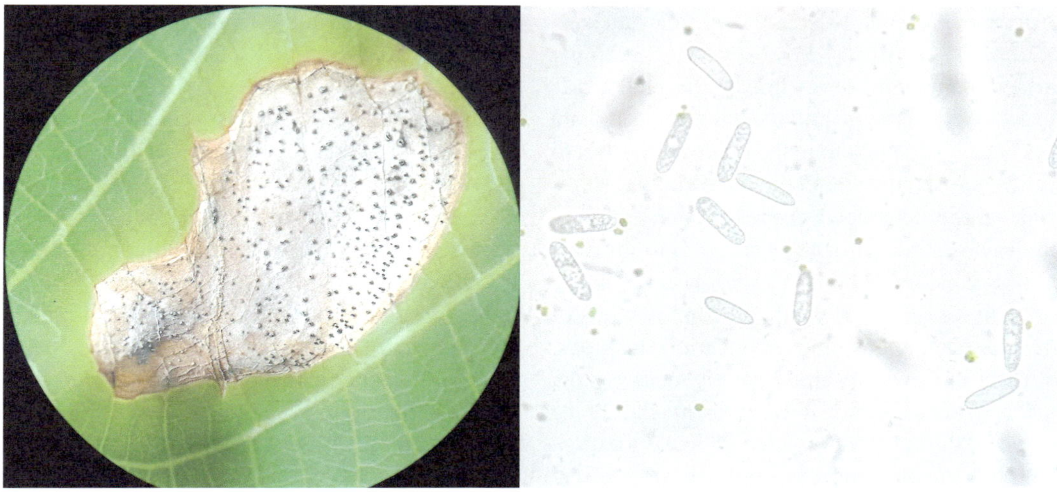

Fig. 12.2 *Colletotrichum*

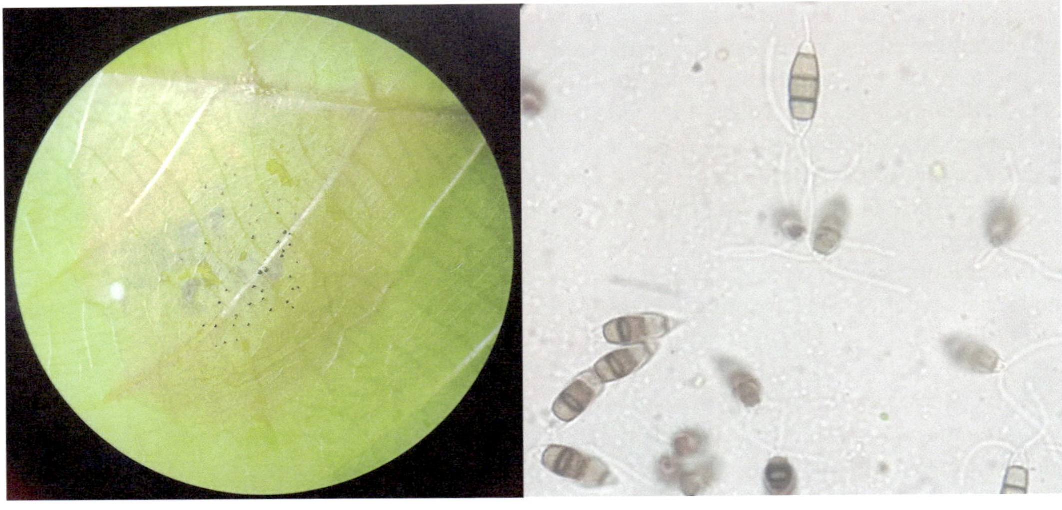

Fig. 12.3 *Neopestalotiopsis*

tures. Second, even within the same genus, the effectiveness of compounds varies when the family differs. Third, there is a lack of field-level pesticide efficacy data. These observations suggest that when using fungicides, it is essential to develop those with high residual activity that are less prone to being washed away by rain. Additionally, to maximize the effectiveness of fungicides, it is crucial to accurately isolate and identify the pathogen at an early stage. By identifying the pathogen, the most effective fungicide or treatment can be selected, avoiding unnecessary use of chemicals. Furthermore, combining multiple fungicides according to the pathogen's life cycle is likely to be important (Okubo-Kurihara et al. 2024). By evaluating the pathogen's susceptibility or resistance, effective management strategies can be developed to prevent the spread of the disease. To control this disease and mitigate its impact on the rubber industry, it is crucial to establish scientifically validated combinations of effective fungicides and application methods, which is an area we are currently working on. Furthermore, pesticide development should be pursued through international collaboration, gathering scientific data to establish more effective fungicides.

12.2 Activities to Protect Natural Rubber Resources from Leaf Fall Disease (LFD)

12.2.1 Outline of SATREPS Project

In 2020, the SATREPS project was started as collaboration between Japan and Indonesia. In Japan, RIKEN Institute (RIKEN) and Gifu University (GU) (later included Maebashi Institute of Technology (MIT)) make up the Japanese team. In Indonesia, Indonesian Rubber Research Institute (IRRI) and Universitas Indonesia (UI) make up the Indonesian team. In RIKEN, two research teams are involved in this project; from The Center for Sustainable Resources (CSRS) is team of Dr. Minami Matsui of the Synthetic Genomics Research group and from the Center for Advanced Photonics (RAP) is the team of Dr. Kiwamu Kase. To achieve the project goals, approaches are taken from immediate, long-term, and preventive perspectives. First, as an immediate response, the project adopts an approach to suppress the current spread of LFD using chemical pesticides and biofungicides. For long-term responses, it provides a fundamental approach aimed at establishing leaf-fall disease-resistant clones through breeding efforts. From a preventive perspective, the project aims to detect infected plantations or rubber trees as early as possible using artificial satellites and drones. This preventive measure also functions as an immediate control approach when combined with chemical pesticides and biofungicides, enabling treatment of infected trees at an early stage to prevent further spread. Lastly, the project emphasizes capacity building for the development of modern breeding techniques for natural rubber trees and the social implementation of technologies developed by SATREPS.

12.2.2 Development of Novel Pesticides and Disease Management

12.2.2.1 Candidate Compound(s) for Novel Fungicide Are Developed for Rubber Tree Leaf Fall Diseases

In this research, we screened fungicides obtained in Japan and in Indonesia. Since Japanese fungicides have never been used to control LFD of natural rubber trees, we firstly obtained possible fungi with same genus with Indonesian LFDs from a Japanese stock center (Genebank project NARO) and examined their growth inhibition by using agar plate assay. In this screening, we obtained two possible fungicides, 8 Hydroquinone copper (Quinondo) and Thiophanate-methyl (Topsin), which inhibit the growth of *Pestalotiopsis* and *Colletotrichum* that are candidates of LFD. (Later UI confirmed natural rubber leaf-fall by *in planta* infection using natural rubber seedlings). In Indonesia, IRRI screened their commercial fungicides in Indonesia. By agar plate assay, they also isolated two LFD suppressive candidates, i.e., Benomil (Benlox) and Mix of Trifloxysutrobin and Tebukonazole (Nativo). Those four fungicides were examined by actual spraying on natural rubber seedlings. Three fungicides, Quinondo, Topsin, and Nativo, suppressed appearance of LFD lesion well (Fig. 12.4). Quinondo is a prophylactic fungicide. It is necessary to be applied before the fungus is infected (Esposito and Fletcher 1961). On the other hand, Topsin has a fungicidal effect. Those two combined with different action timing will have broad efficacy against LFDs.

We next examined a natural rubber budwood garden in IRRI for prevention of LFDs in natural conditions. In this experiment, we examined fungicides prevention effects against LFD in natural environments using several rubber clones (Fig. 12.5). Although there are some differences about suppression degree, three fungicides, i.e., Quinonodo, Topsin, and Nativo, suppressed lesion spots formation by LFDs efficiently. In natural environments, plant condition changes frequently. It is necessary to optimize suitable spraying timing and intervals and fungicide concentration.

12.2.2.2 Candidate Microbial Formulations for Novel BiofungicideAreDeveloped for Rubber Tree Leaf Fall Diseases

Our research collaboration focuses on developing biological solutions to combat Leaf Fall Disease (LFD). While previous studies have explored microbial biocontrol agents (MBCAs) against fungal pathogens such as *Colletotrichum* spp. and *Corynespora cassiicola* (Zhang et al. 2020; Mahendran et al. 2021; Seekham et al. 2022; Wang et al. 2024), none of the studies have reported the MBCAs against *Pestalotiopsis*-related species, which are major contributors to LFD in Indonesia.

To bridge this gap, our project brings together expertise from GU in Japan and the IRRI in Indonesia. By combining advanced microbial research with practical field knowledge, we aim to develop a novel biofungicide using MBCA specifically targeting these understudied pathogens.

The project initiated with an extensive bacterial isolation process, meticulously collecting and cataloging over 400 and 60 strains of symbiotic bacteria from rubber leaves in Japan and Indonesia, respectively. This comprehensive collection provided a robust foundation for screening potential biocontrol candidates.

The initial screening used a detached leaf assay, a critical method for evaluating potential

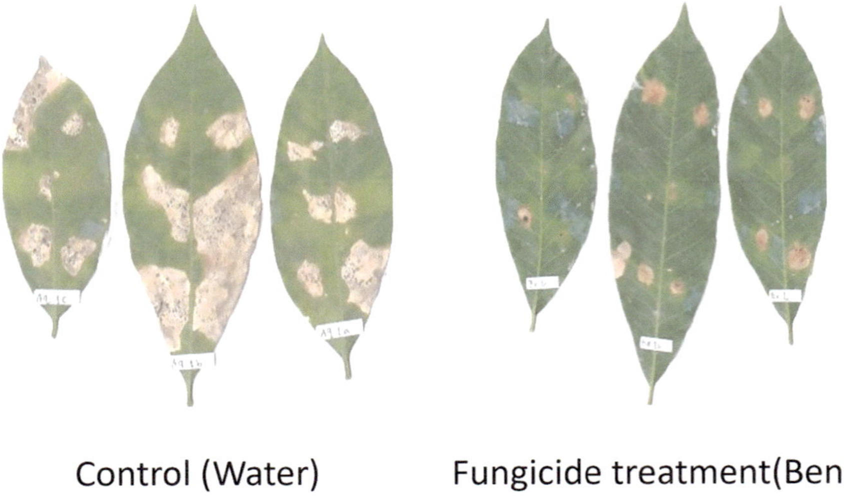

Fig. 12.4 Inhibition of fungal growth by chemical fungicide (Benlox). Fungicide was sprayed before fungal (*Neopestalotiopsis* sp. isolated from GT1) inoculation on GT1 leaves. These leaves were wounded at six points with a syringe needle and inoculated with mycelial plugs (six points in each leaf. In left three leaves, sprayed water as control; in right three leaves, sprayed fungicide)

Fig. 12.5 IRRI Budwood Garden. Rubber clones BPM24, PR303, PB260, PB340, and dIRR39 were grown. Each black square contains 50 trees. Four fungicides Quinondo, Topsin, Dithane, Nativo, and water (control) were sprayed for 1, 2, 3, 4, 5 squares, respectively

biocontrol candidates. In this experimental approach, rubber leaves were pre-treated with individual bacterial strains and then challenged with *Pestalotiopsis microspora* and *Colletotrichum siamense*, allowing us to systematically assess inhibitory effects against the development of necrotic lesions (Fig. 12.6).

The screening process identified five promising bacterial strains with remarkable disease suppression potential. However, rigorous safety evaluations were critical because harmful microbes are not acceptable for use in biocontrol. A sheep blood agar test revealed hemolytic activity in three of the strains, indicating potential risks as opportunistic pathogens for humans and animals, and the remaining two strains were selected for further evaluation.

These selected strains had several key characteristics essential for ideal biocontrol agents. For example, they exhibited high tolerance to extreme environmental conditions, particularly heat and drought, through their ability to produce endospores—a critical factor in developing a practical biofungicide.

Detailed in vitro experiments unveiled the strains' remarkable capabilities. They demonstrated the ability to produce both lytic enzymes and antifungal metabolites with potent inhibitory activity against LFD pathogens. Microscopic observations provided additional crucial insights: pretreatment with these bacterial strains significantly inhibited spore germination of *P. microspora* and *C. siamense* on rubber leaves (Fig. 12.7). By interfering with spore germination, the bacterial strains showed potential to interrupt disease progression at its earliest stage.

Fig. 12.6 Efficacy of the candidate strains on detached rubber leaves (unpublished data). Detached rubber leaves were pretreated with bacterial suspension and incubated for 24 h prior to pathogen inoculation

Recognizing the importance of real-world validation, the research team secured official approval from the Indonesian Ministry of Agriculture to import and initiate further experiments with the candidate strains in Indonesia. Field trials are currently underway at IRRI, monitoring the effectiveness of these potential biocontrol agents under actual rubber plantation conditions.

The project aims to develop a commercially viable biofungicide in partnership with an Indonesian biopesticide company, with upcoming steps including formulation finalization and pesticide registration. By bridging fundamental microbiological research with practical agricultural applications, the project offers a promising and eco-friendly approach to addressing a critical challenge in rubber production.

12.2.3 Breeding and Genetic Improvement for Disease Resistance in *Hevea brasiliensis*

12.2.3.1 Novel Clones of *Hevea brasiliensis* That DemonstratestheResistance AgainstRubberTreeLeafFall Diseases Are Produced

The main objective of this output is to obtain novel clones of *Hevea brasiliensis* that are resistant to *Pestalotiopsis* leaf fall disease. In this study, we divided the research goals into three categories: determining the procedure for leaf fall disease selection of rubber clones based on genome analysis, identifying markers to define the leaf fall disease resistance of rubber clones, and promoting crossbreeding using the markers identified for producing a novel resistance rubber clone.

A consortium comprising IRRI, RIKEN, MIT, UI, and GU is conducting research activities to achieve the outlined goals. IRRI provided genetic resources and phenotyping traits, including F1 progenies from the cross between PB260 and RRIC100, field selection of these F1 progenies focusing on growth, latex potential yield, and dis-

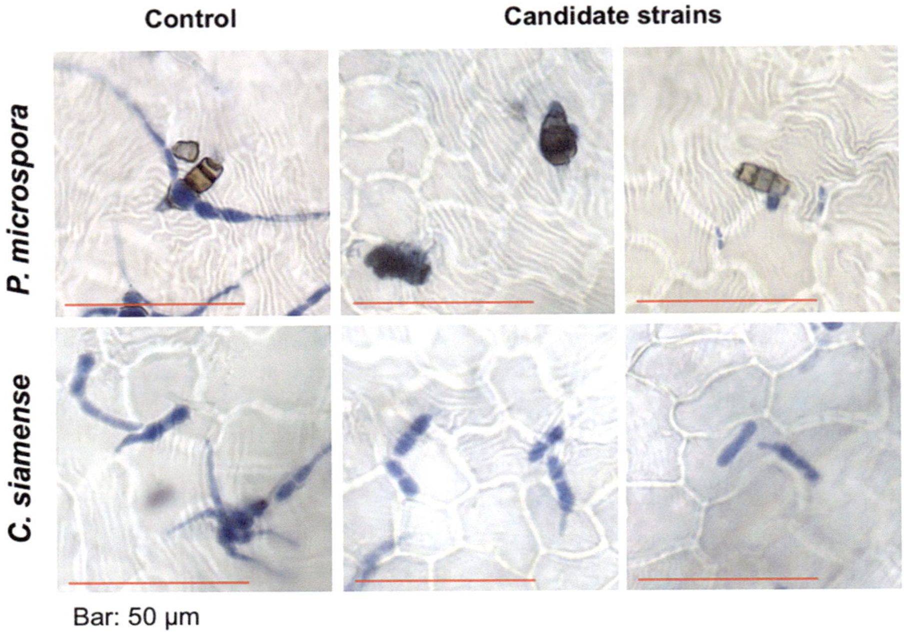

Fig. 12.7 Inhibition of the spore germination of the pathogens, *P. microspora* and *C. siamense*, by the selected bacterial strains on rubber leaves (unpublished data)

ease resistance, adhering to Good Agricultural Practice (GAP). RIKEN and MIT performed further work, including genome sequencing, Genome-Wide Association Studies (GWAS), and genome selection for breeding disease-resistant varieties. In addition, UI and GU cooperatively determined the experimental procedures to extract RNA and determined the marker gene for the LFD infection from the RNA-sequencing data.

To elucidate the mechanisms of LFD infection, UI and GU obtained the transcriptome analysis of infected leaf samples. The focus is on identifying markers that are consistently differentially expressed across all time points, offering insights into the plant's response to infection and its resistance mechanisms. The RNA-sequencing data can serve as a basis for developing marker databases correlated with resistance to *Pestalotiopsis*, aiding in the future development of more tolerant rubber clones for breeding programs.

Artificial inoculation of fungi for RNA-sequencing was conducted at both UI and GU with the GT1 rubber clone as a model (Fig. 12.8). The whole plant assay was performed at UI using a budded stump, while GU employed the detached leaf assay. The inoculation method for both assays was consistent, involving the use of *Pestalotiopsis* isolates on agar plugs with prior wounding, with a mock treatment serving as a control. Each treatment includes three biological replicates to ensure reliable symptom assessments were observed at 2, 4, and 6 days post-inoculation. These time points were strategically selected to capture the progression of the disease and the pathogen's impact, providing valuable insights into the temporal dynamics of infection and resistance mechanisms.

The whole plant assay facilitates the observation of rubber leaf fall from the first day of inoculation, helping to establish the timeline of *Pestalotiopsis*-induced leaf fall. Using this *in planta* approach with budded stump, the natural rubber leaf-fall phenomenon has been effectively confirmed. It also considers the entire plant's physiological responses, including systemic resistance mechanisms. According to Miller-

Fig. 12.8 Artificial inoculation methods

Butler et al. (2018), the detached leaf assay can also reliably and reproducibly identify plant resistance, similar to the whole plant assay.

Overall, this study demonstrates that both assays can be effectively used to investigate pathogen phenotypic diversity. The whole plant assay provides a comprehensive understanding of plant resistance, while the detached leaf assay offers a quicker and more cost-effective alternative for initial screening. By combining the insights from both methods, alongside the application of RNA sequencing, the accuracy of identifying resistant rubber plant varieties to combat leaf fall disease caused by *Pestalotiopsis* can be enhanced.

12.2.3.2 Genome-Oriented Technology for Developing New Clones in Rubber Tree

Breeding strategies for *Hevea brasiliensis* (rubber tree) aim to enhance latex yield, disease resistance, and environmental adaptability, addressing challenges posed by plant diseases and climate change. This chapter outlines strategies for developing new clones. As traditional breeding, selection breeding has been widely used in rubber-producing countries. Its advantages include simplicity and low cost, but a major drawback is the gradual loss of genetic diversity, limiting long-term improvements in yield and resistance.

Genome-wide association studies (GWAS) are a powerful tool in modern plant breeding, identifying genetic loci associated with critical traits. This method comprehensively searches for correlations between phenotypic polymorphisms and DNA polymorphisms (genotypes) across the genome, revealing genetic mechanisms linked to desired traits.

In the SATREPS project, we used our genome data of RRIM 600, a major high-yielding clone used in Malaysia and Thailand. GWAS analysis has not been performed on *Pestalotiopsis* leaf fall disease, our main goal. For this purpose, we generated over 500 F1 progeny by crossing RRIC 100, a clone tolerant to LFD pathogens *Colletotrichum* and *Corynespora*, with PB 260, a high latex yielding clone. Over 300 F1 progenies were sequenced, and their SNPs and STRs were identified across the entire genome. Both field observations against *Pestalotiopsis* leaf fall disease and detached leaf assays were conducted for phenotypic data analysis. We also performed GWAS analysis from field observation data (Fig. 12.9). We could observe the highest peak in chromosome 11. Based on these genomic loci, we will examine the relationship between LFD-resistance phenotype and corresponding gene function (Makita et al. 2024).

12.2.4 A Detection System for Rubber Tree Disease-Affected Areas by AI Imaging Analysis of Rubber Plantations Is Developed Using Data Acquired from Artificial Satellites and Drones

To prevent further spread of the LFD, it is necessary to quickly identify new infected areas and treat them with chemical and biofungicides. To this end, we are developing a system to regularly observe rubber fields and monitor changes (Fig. 12.10).

We have constructed a WEB site of IRRI Sembawa plantation as a model for detection of LFD. We created training data from highly accurate spectral images of rubber tree leaves and canopy and from rubber tree vegetation data.

The RIKEN RAP, Geography, and Mathematics of UI generated training data based on infection status from satellite images and drone images of IRRI Sembawa plantation. In addition, to obtain local infection status that cannot be captured at the macrolevel, photos of rubber tree affected by LFD and location information by GPS are being collected from rubber farmers as collaboration.

LFD-infected rubber leaves images are sent to UI and UI mathematics members classified them according with the predetermined infection degrees. By repeating this classification, accuracy of detection of LFD-infected leaves and area will improve. The obtained local infection information will also be provided to farmers through the Internet, enabling them to understand the status of LFD infection in their neighborhood with geographical maps (Fig. 12.11).

12.2.5 Capacity Building for Development of Modern Breeding of Natural Rubber Tree and Social Implementation of Technologies Developed by SATREPS

In this SATREPS, we expect that Indonesian rubber researchers can incorporate modern technologies for breeding and control quality of natural rubber. Currently three Ph.D. candidates are studying at GU through the support of the MEXT international graduate student program. From 2022, we started a short visiting course for

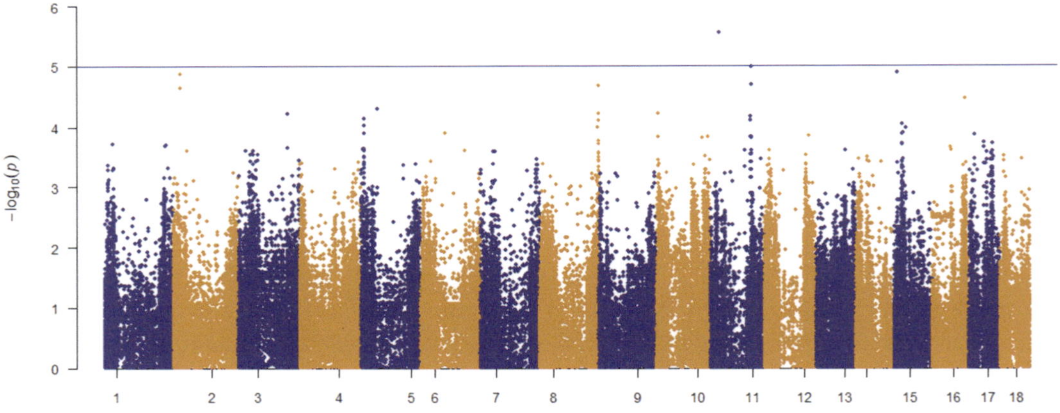

Fig. 12.9 GWAS analysis of Pestalotiopsis leaf fall disease (Makita et al. 2024)

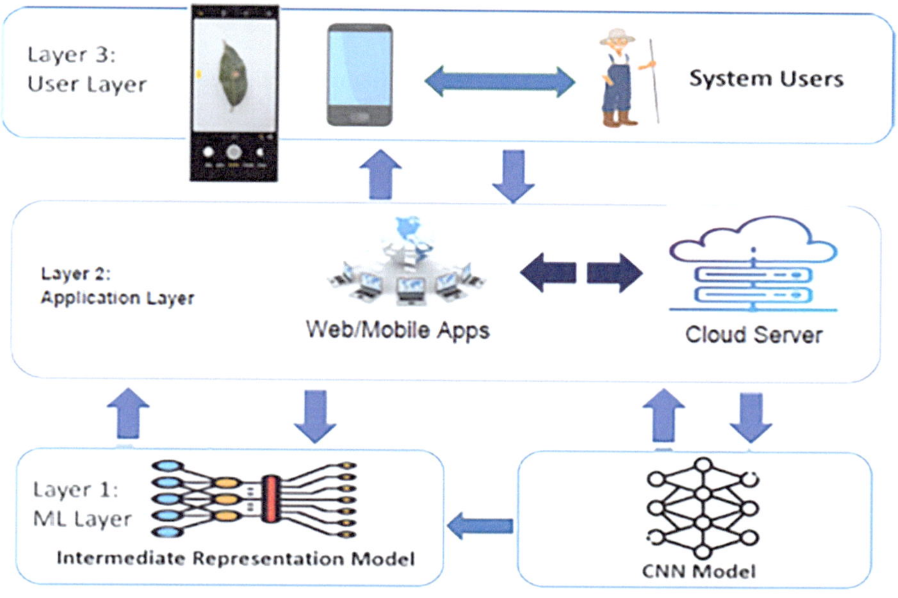

Fig. 12.10 Three layers for implementation of early detection of LFD-infected areaLayer 3: Images of leaves affected by LFD taken by farmers are sent to the UI with GPS informationLayer 2: The UI classifies the leaf data collected from farmers using AILayer 1: Supervised data are generated from training image data to improve the accuracy of classification of infection degree

Fig. 12.11 Satellite imaging for monitoring plantation, left: Monitoring area, right: NDVI (Normalized Difference Vegetation Index) with date (from 2019 to 2022). NDVI represents photosynthetic activities of leaves. In 2019, NDVI reduces in May to June indicating seasonal leaf fall. In 2020, leaf fall continues from May to September. In 2021, a heavy reduction of NDVI was observed. In 2022, a similar heavy NDVI reduction was observed from January to March and also several unusual NDVI reduction can be observe with leaf-fall by LFD

Indonesian researchers every year. In 2024, 10 Indonesian researchers were visited to Japan for 2 weeks to have training of bioinformatics at MIT, molecular biology in RIKEN, and plant pathology in GU. To explore research results and discussion, IRRI organized international rubber conference in Indonesia in 2024. Around 300 researchers from most rubber producing countries were gathered to have presentation including leaf-fall disease.

12.3 Future Outlook for Natural Rubber

12.3.1 Future of Natural Rubber as a Sole Resource of World Transportation

As previously stated, natural rubber has been a cornerstone of various global industries for centuries. Due to its unique properties of elasticity, resilience, and resistance, it is crucial for plenty of usages. Industries such as medtech, consumer goods, and transportation are some of the main consumers of natural rubber. Vital necessities such as surgical gloves, catheters, seals, and other medical equipment harnessed the indispensable biocompatibility and properties of natural rubber. Natural rubber is also an indispensable commodity that carries the entirety of tire industry. In civil engineering, natural rubber is utilized to create viscoelastic dampers in anti-earthquake buildings (Ramakrishna and Mohan 2020) and as asphalt modifier and bio-modifier for pavement construction, opening a whole new option of environmentally friendly engineering practices (Ansari et al. 2021).

With such tremendous utilities, natural rubber production is important to sustain related industry. However, as the world transitions towards a more sustainable future, the natural rubber industry is poised for significant transformation due to challenges. One of the major concerns is the environmental impact of rubber cultivation, particularly deforestation (Jayathilake et al. 2024). Additionally, the production of natural rubber is vulnerable to factors such as climate change and disease outbreaks. For example, the natural rubber industry, especially in the Southeast Asia region, is facing a massive outbreak of the novel *Pestalotiopsis* Leaf Fall Disease (PLFD) (Febbiyanti and Fairuzah 2020). The outbreak has caused production loss since 2018 and needs to be immediately resolved. Thus, the rubber industry is exploring various strategies to address those challenges.

12.3.2 Practicing Circular Economy

Practicing circular economy induces a sustainable supply chain and impactful to the environment. Reclaiming and reprocessing latex scrap, creating products from recycled rubber, and reusing end-of-life rubber products are some of the activities that can be initiated. For one, recycled rubber can be used in applications such as road surfacing, construction materials, and even athletic tracks (Hassan and Rodrigue 2024). By adopting circular economy principles, the natural rubber sector aims to create a more sustainable and resource-efficient value chain.

12.3.3 Implementing Rubber Agroforestry as Sustainable Agriculture Practice

As reported by Zou et al. (2024), the Integrated Rubber Agroforestry System (IRAS) can potentially play a role in preserving biodiversity, improving soil structure and fertility, and positively impacting the ecosystem. IRAS increases land use efficiency and contributes to socioeconomic improvement for smallholders, especially in rural areas. Agroforestry also enables us to produce specific commodities while also offering ecosystem services like carbon sequestration, water regulation, and biodiversity habitat provision through the strategic selection of compatible species and management techniques. These are great benefits that can also reduce the negative impact of deforestation due to the current plantation model.

12.3.4 Resource Diversification

Alternative bioresources such as Guayule (*Parthenium argentatum*) and Russian dandelion (*Taraxacum kok-saghyz*) can be utilized to help fulfill natural rubber demand, which mostly covered by the para-rubber trees (*Hevea brasiliensis*) (Kohjiya and Ikeda 2014). Natural rubber and latex from both sources have very similar composition and performance to that from *H. brasiliensis*. Guayule and Russian dandelion are especially prospective in countries where para rubber trees cannot thrive well, like North America, Europe, and East Asia.

Aside from addressing current challenges, it is also important to pay attention to the intrinsic value of the natural rubber industry. First of all, the para rubber tree plantation is proven to be a potential carbon sequestration site. According to Maggiotto et al. (2014), a 15-year-old rubber plantation can store up to 79.3 mg C ha^{-1} of carbon in the top 60 cm of the soil. In another report from Lestari and Noor'an (2022), the rubber plantation in East Kalimantan was able to store up to 75.71 tons C ha^{-1} of carbon. Secondly, the natural rubber's unique properties are almost impossible to replicate. There are claims saying that synthetic rubber offers greater consistency, lower cost, and better resistance to certain external factors. Unfortunately, synthetic rubber lacks some of the unique properties of natural rubber, rendering it subpar in terms of performance. Another downside of synthetic rubber is the materials it is derived from—petroleum-based chemicals. The nature of its production is extremely concerning for the environment, making it obvious that synthetic rubber may not be the best alternative to natural rubber.

Based on the provided elaborations, it is reasonable to consider natural rubber as an indispensable resource for many industries. However, immediate actions need to be taken in order to face challenges awaiting the natural rubber production scene. Switching current approaches to something more sustainable, such as implementing circular economy, sustainable agriculture, and resource diversification, is a viable option. Accompanied with the intrinsic value of the natural rubber industry, natural rubber is surely securing its place as the most significant, if not the sole, resource of world traffic.

References

Aliya SSS, Nusaibah SA, Mahyudin MM, et al (2022) Colletotrichum siamense and Pestalotiopsis jesteri as potential pathogens of new rubber leaf spot disease via detached leaf assay. J Rubber Res 25:195–212. https://doi.org/10.1007/s42464-022-00157-4

Ansari AH, Jakarni FM, Muniandy R, Hassim S (2021) A review on the application of natural rubber as Asphalt Modifier. IOP Conf Ser Mater Sci Eng 1075(1):012031. https://doi.org/10.1088/1757-899x/1075/1/012031

Chen L, Xu L, Li X, Wang Y, Feng Y, Huang G (2023) The diseases and pests of rubber tree and their natural control potential: a bibliometric analysis. Agronomy 13:1965. https://doi.org/10.3390/agronomy13081965

Damiri N, Pratama YP, Febbiyanti TR, Rahim SR, Astuti DA, Purwanti Y (2022) Pestalotiopsis sp. infection causes leaf fall disease of new arrivals in several clones of rubber plants. Biodivers J Biol Divers 23:3943–3949

Darojat MR, Ardhie SW, Oktavia F, Sudarsono. (2023) New leaf fall disease in rubber: pathogen characterization and rubber clone resistance evaluation using detached leaf assay. Biodiversitas 24:1935–1945

Esposito RG, Fletcher AM (1961) The relationship of pteridine biosynthesis to the action of copper 8-hydroxyquinolate on fungal spores. Arch Biochem Biophys 93:369–376

FAO. FAOSTAT. Accessed 23 Nov 2023. https://openknowledge.fao.org/items/1516eb79-8b43-400e-b3cb-130fd70853b0

FAO. FAOSTAT. Accessed 5 Aug 2024. https://www.fao.org/faostat/en/#home

Febbiyanti TR, Fairuza Z (2019) Identification of causes of rubber leaves outbreak in Indonesia. Indo J Nat Rubber Res 37(2):193–206. https://doi.org/10.22302/ppk.jpk.v37i2.616

Febbiyanti TR, Fairuzah Z (2020) Identifikasi Penyebab Kejadian Luar Biasa Penyakit Daun Karet di Indonesia. Jurnal Penelitian Karet:193–206. https://doi.org/10.22302/ppk.jpk.v37i2.616

Guerra NB, Pegorin GS, Boratto MH, Barros NR, Graeff CF, Herculano RD (2021) Biomedical applications of natural rubber latex from the rubber tree *Hevea brasiliensis*. Mater Sci Eng C Mater Biol Appl 126:112–126

Guyot J, Omanda E, Pinard F (2005) Some epidemiological investigations on *Colletotrichum* leaf disease on rubber tree. Crop Prot 24:65–77. https://doi.org/10.1016/J.CROPRO.2004.06.009

Hassan MR, Rodrigue D (2024) Application of waste tire in construction: a road towards sustainability and circular economy. Sustain For 16(9):3852. https://doi.org/10.3390/su16093852

Herath K, Samita S, Wijesuriya W (2012) Identification of common trends in prices, demand and supply of natural rubber: an approach using dynamic factor model. Trop Agric Res 24:54–70

Hunupolagama DM, Chandrasekharan NV, Wijesundera WSS, Kathriarachchi HS, Fernando THPS, Wijesundera RLC (2017) Unveiling members of *Colletotrichum acutatum* species complex causing *Colletotrichum* leaf disease of *Hevea brasiliensis* in Sri Lanka. Curr Microbiol 74(6):747–756. https://doi.org/10.1007/s00284-017-1238-6

IRC (2024) Accessed 19 Dec 2024. https://www.puslitkaret.co.id/welcome-irc-2024/

Jayasinghe CK, Fernando TH, Priyanka UM (1997) Colletotrichum acutatum is the main cause of *Colletotrichum* leaf disease of rubber in Sri Lanka. Mycopathologia 137(1):53–56. https://doi.org/10.1023/A:1006850119146

Jayathilake HM, Wei CZ, Tan G, De La Porte C, Carrasco LR (2024) Rubber plantation size and global rubber price are linked to forest loss and degradation in Jambi, Sumatra. Intgr Conserv 3(1):22–35. https://doi.org/10.1002/inc3.43

Kohjiya S, Ikeda Y (eds) (2014) Chemistry, manufacture and applications of natural rubber. Elsevier. https://doi.org/10.1016/c2013-0-16344-1

Kusdiana APJ, Saputra J (2022) Impact of *Pestalotiopsis* leaf fall disease on leaf area index and rubber plant plantations. IOP Conf Ser Earth Environ Sci 995(1):012030. https://doi.org/10.1088/1755-1315/995/1/012030

Kusdiana APJ, Sinaga MS, Tondok ET (2020) Diagnosis of causal agent of new rubber leaf fall disease (*Hevea brasiliensis* Muel. Arg). Indo J Nat Rubber Res 38(2):165–178

Lestari NS, Noor'an RF (2022) Carbon sequestration potential of rubber plantation in East Kalimantan. IOP Conference Series: Earth and Environmental Science 1109(1):012102. https://doi.org/10.1088/1755-1315/1109/1/012102

Lieberei R (2007) South American leaf blight of the rubber tree (*Hevea* spp.): new steps in plant domestication using physiological features and molecular markers. Ann Bot 100(6):1125–1142. https://doi.org/10.1093/aob/mcm133

Li B, Liu X, Cai J, Feng Y, Huang G (2021) First report on Neopestalotiopsis aotearoa of rubber tree in China. Plant Dis 105(4):1223. https://doi.org/10.1094/PDIS-09-20-1930-PDN

Maggiotto SR, De Oliveira D, Marur CJ, Stivari SMS, Leclerc M, Wagner-Riddle C (2014) Potential carbon sequestration in rubber tree plantations in the Northwestern region of the Paraná State, Brazil. Acta Sci Agron 36(2):239. https://doi.org/10.4025/actasciagron.v36i2.17404

Mahendran TR, Thottathil GP, Surendran A, Nagao H, Sudesh K (2021) Biocontrol potential of *Aspergillus terreus*, Endophytic fungus against *Rigidoporus microporus* and *Corynespora cassiicola*, pathogens of rubber tree. Arch Phytopathol Plant Protect 54:1014–1032

Makita Y et al (2024) Constructing of genome-based infrastructure for developing resistant clones against *Hevea* leaf fall disease. International Rubber Conference

Mazlan NA, Harikrishna JA, Mohamed R (2019) Advances in rubber tree genomics and Transcriptomics. Rubber Res 22(1):1–15. https://doi.org/10.1007/s42464-019-00007-1

Miller-Butler MA, Smith BJ, Babiker EM, Kreiser BR, Blythe EK (2018) Comparison of whole plant and detached leaf screening techniques for identifying anthracnose resistance in strawberry plants. Plant Dis 102(11):2112–2119. https://doi.org/10.1094/PDIS-08-17-1138-RE

Okubo-Kurihara E, Febbiyanti TR, Ashari Y, Yanagawa EO, Kuriyama T, Shimizu M, Diyasti F, Matsui M (2024) Screening of effective pesticides to control rubber tree leaf fall disease (LFD) caused by *Neopestalotiopsis* and *Colletotrichum* fungi in Indonesia. J Pestic Sci 49:277–284. https://doi.org/10.1584/jpestics.D24-020

Pornsuriya P, Sangpradit K, Yutthachai T, Kangsadalampai K (2020) Rubber tree diseases caused by *Colletotrichum*: symptoms and management strategies. Thai J Agric Sci 53(3):145–158

Premadasa AD (2020) Decline in natural rubber production: analysis and mitigation strategies. J Nat Rubber Res 33:113–128

Priyadarshan PM (2017) Biology and agronomy of rubber. Elsevier Science

Ramakrishna U, Mohan S (2020) Performance of low-cost viscoelastic damper for coupling adjacent structures subjected to dynamic loads. Mater Today Proc 28:1024–1029. https://doi.org/10.1016/j.matpr.2019.12.343

Seekham N, Kaewsalong N, Jantasorn A, Dethoup T (2022) Biological control of *Corynespora* leaf fall disease in rubber by Endophytic *Trichoderma* spp. under field conditions. Eur J Plant Pathol 170:207–218

Statistics Indonesia. Indonesian Rubber Statistics 2022. Accessed 5 Aug 2024. https://www.bps.go.id/id/publication/2023/11/30/8cdca0a6a45235c12ed4c4d1/statistik-karet-indonesia-2022.html

Syarifa LF, Agustina DS, Alamsyah A, Nugraha IS, Asywadi H (2023) Commodity outlook of Indonesian natural rubber 2023. Indo J Nat Rubber Res 41(1):47–58

Thaochan N, Wongkaew S, Singh T (2020) Efficacy of Carbendazim against *Neopestalotiopsis cubana*: a promising solution for rubber leaf fall disease. Plant Pathol J 36(4):412–419. https://doi.org/10.5423/PPJ.OA.12.2019.0298

Vineeth TV, Rajesh R, Anilkumar A (2024) The resurgence of *Colletotrichum* leaf disease in Kerala rubber plantations: causes and solutions. Indian J Plant Pathol 42(2):203–215. https://doi.org/10.1007/s11256-024-1234-6

Wang M, Zhang Y, Cai H, Zhao X, Zhu Z, Yan Y, Yin K, Cheng G, Li Y, Chen G, Zou L, Tu M (2024) A new biocontrol agent *Bacillus velezensis* SF334 against rubber tree fungal leaf anthracnose and its genome analysis of versatile plant probiotic traits. J Fungi 10(2):158. https://doi.org/10.3390/jof10020158

Zhang K, Gu L, Zhang Y, Liu Z, Li X (2020) Dinactin from a new producer, *Streptomyces badius* gz-8, and its antifungal activity against the rubber anthracnose fungus *Colletotrichum gloeosporioides*. Microbiol Res 240. https://doi.org/10.1016/j.micres.2020.126548

Zou R, Sultan H, Muse Muhamed S, Khan MN, Pan J, Liao W, Li Q, Cheng S, Tian J, Cao Z, Tao Y, Nie L (2024) Sustainable integration of rubber plantations within agroforestry systems in China: current research and future directions. Plant Sci Today 11(3). https://doi.org/10.14719/pst.2024.11.3.4180

Role and Perspectives of Plant Genetic Resources for Guaranteeing Sustainable Crop Production in Developing Countries

13

Kenji Irie, Tsukasa Nagamine, and Misa Masuda

Abstract

Plant genetic resources are essential for sustaining global food production and addressing agricultural challenges such as climate change, pest outbreaks, and declining biodiversity. The rich biodiversity of tropical and subtropical regions has historically contributed to the development of diverse crop varieties through natural and artificial selection. These genetic resources hold valuable traits, such as pest resistance and environmental stress tolerance, which are critical for developing superior crop varieties. However, the sustainable use and conservation of these resources require international cooperation and compliance with frameworks like the Nagoya Protocol and the International Treaty on Plant Genetic Resources for Food and Agriculture. SATREPS have facilitated research, capacity building, and international collaboration to promote the sustainable use of plant genetic resources. Agricultural biodiversity, encompassing cultivated crops, wild species, and ecological functions, plays a vital role in food security, ecosystem resilience, and nutrition. Diversifying agricultural systems and diets can address global issues like "hidden hunger" and malnutrition while supporting sustainable and resilient food systems. Strengthening global networks, conserving genetic diversity, and fostering equitable benefit-sharing mechanisms are crucial for ensuring the availability of genetic resources for future generations and achieving sustainable agricultural development.

Keywords

Plant genetic resources · Sustainability · Food security

13.1 The Role of Plant Genetic Resources

The total number of species on Earth is estimated to be between 5 and 30 million, with more than half of these species found in tropical and subtropical regions. Tropical and subtropical regions are rich in biodiversity, and throughout human history, we have used a wide variety of organisms, from animals and plants to microorganisms, for food, clothing, housing, medicine, and industrial materials. Organisms useful to humans

K. Irie (✉)
Tokyo University of Agriculture,
Setagaya, Tokyo, Japan
e-mail: k3irie@nodai.ac.jp

T. Nagamine
Ex-National Agriculture and Food Research Organization (NARO), Tsukuba, Ibaraki, Japan

M. Masuda
University of Tsukuba, Tsukuba, Ibaraki, Japan

T. Nagamine et al. (eds.), *Plant Genetic Resources (PGR) for Sustainable Crop Production*, Sustainable Development Goals Series, https://doi.org/10.1007/978-981-96-7117-5_13

are a gift of nature, discovered by our ancestors over tens of thousands of years, and there is a need to conserve and sustainably use the genetic resources of organisms useful to our lives. The diverse crops that our ancestors have cultivated over thousands of years have spread from their original regions to all parts of the world through interactions and migrations between ethnic groups. In the process, they have adapted to the different natural environments of each region, including latitude, day length, topography, temperature, rainfall, water availability, and soil, and have differentiated into a wide range of varieties against the backdrop of the different food cultures and eating habits of each ethnic group. These diverse varieties, the product of natural and artificial selection, form the basis of today's agricultural genetic resources. These crop varieties are likely to contain useful genes related to resistance to pests and diseases, tolerance to poor environments and other traits, as well as traits related to yield and quality. In recent years, global warming caused by climate change has led to problems such as drought, damage from high temperatures, and the spread of new pests and diseases and has also affected stable agricultural production. To address these problems, it is necessary to promote the development of new varieties, and plant genetic resources with diverse genetic traits are essential for this purpose. Plant genetic resources play an important role in the development of superior varieties. If there are genetic resources with genetic variation that can be used to improve crops, there is a chance that the desired gene can be found. For this reason, it is important to conserve a wide range of genetic variation and genetic diversity.

13.2 Sustainable Use of Plant Genetic Resources

While the importance of conserving plant genetic resources is widely recognized throughout the world, developing countries in tropical and subtropical regions that are rich in genetic resources have a concept of resource nationalism, which involves managing and developing their own genetic resources in their own countries. In this context, fair and equitable benefit-sharing with resource-rich countries is needed to access genetic resources from abroad and use them for academic research, commercial products, or as breeding material for new varieties. The International Treaty on Plant Genetic Resources for Food and Agriculture, which entered into force in 2004, promotes the conservation and sustainable use of plant genetic resources for food and agriculture and the fair and equitable sharing of the benefits arising from their use. This means that it is now possible to obtain agricultural plant genetic resources through the Multilateral System (MLS). At the 10th Conference of the Parties to the Convention on Biological Diversity (COP10) held in Nagoya in 2010, the Nagoya Protocol on Access to Genetic Resources and the Fair and Equitable Sharing of Benefits Arising from their Utilization (ABS) was adopted. In 2014, the Protocol entered into force after being ratified by 50 countries, and from now on, the strict rules of the Nagoya Protocol must be observed by the provider country when accessing and using genetic resources from abroad.

In the context of global climate and environmental changes, the importance of research and development related to the production, use, and management of biological resources, especially in developing countries, has been emphasized to ensure that humanity will continue to enjoy the benefits of biological resources in the future. At the outset of the SATREPS program, when project participants obtain genetic resources and use them for research and development, they must review and comply with the laws and regulations of the country of origin and work to obtain PIC and MAT. In the field of SATREPS biological resources, 47 research projects have been implemented so far, and this book focuses on 11 research projects related to plant genetic resources that have been implemented in recent years. Some of these projects have achieved results using genetic resources, which are one of the basic services of biodiversity. In India, the biological nitrification inhibition (BNI) trait of the closely related wild species Leymus racemosus was introduced into wheat varieties, and wheat lines with excellent nitrogen-use efficiency were developed. In Madagascar, a new cultivar with the *Pup1* gene locus, which promotes root development, has been developed, and similar results using genetic resources are expected from ongoing projects.

13.3 Building a Network for the Sustainable Use of Genetic Resources

As awareness of rights over genetic resources has increased in developing countries in recent years, and the introduction of new plant genetic resources from overseas has become more difficult, close and extensive international cooperation plays an important role in promoting the conservation and use of plant genetic resources beyond national and regional frameworks. It is essential for developed countries, which are short of genetic resources, to strengthen their activities in the field of plant genetic resources at the international level by providing relevant technologies for the conservation and use of genetic resources, exchanging information, and promoting joint research (FAO, 2010). In Japan, national and public research institutes, international cooperation organizations, universities, and private companies have each conducted joint research and collaborative projects. In the field of biological resources, under the SATREPS program, we have also worked to develop human resources in partner countries in the field of plant genetic resources through international research and development. The development of international networks through human resource development will promote the smooth introduction and use of genetic resources.

13.4 Agricultural Biodiversity for the Future of Food

Agricultural biodiversity encompasses a wide range of elements within food systems, including cultivated plants and farmed animals, wild foods, and other products collected by rural communities in traditional subsistence systems. It also includes organisms such as pollinators and soil biota that are essential in production environments. Agricultural biodiversity contributes significantly to food systems and global sustainability by providing:

1. A source of genetic resources for agriculture
2. Pest and disease control through natural ecological mechanisms
3. Climate risk reduction, enhancing resilience to environmental changes
4. Improved nutrition, ensuring diverse and balanced diets
5. Ecosystem maintenance, such as soil conservation and pollinator support
6. Livelihood opportunities, preserving traditional knowledge and cultural practices

These ecological functions collectively support diverse and sustainable food systems that feed the world.

Globally, much of our food energy intake comes from a few highly productive crops, such as maize, wheat, and rice, along with sugar, processed fats, and oils. While these crops have ensured food security, this reliance has led to "Hidden Hunger," a condition where people lack essential vitamins and trace elements. Over 2 billion people worldwide are estimated to suffer from hidden hunger.

To combat nutritional challenges, we need to diversify our diets by introducing a variety of nutrient-rich crops and vegetables. By promoting agricultural biodiversity and diversifying agricultural products, we can improve the nutritional value of food consumption.

Agricultural biodiversity offers solutions to malnutrition, obesity, and other health issues in both developing and developed countries. It provides essential micronutrients, vitamins, and functional components directly through diverse food resources. Traditional crops, often overlooked, contain a wide array of nutritional and functional compounds (Keding and Cogill 2013). Additionally, agricultural biodiversity enables farmers to develop improved crop varieties that are more resistant to climate change, pests, and diseases, while addressing nutritional deficiencies. To ensure food security in the future, it is essential to develop a system that balances quantity (adequate food supply) with quality (ensuring health and preventing obesity and disease) (Johns et al. 2006). Agricultural biodiversity is recognized as a production system that supports the health of ecosystems, communities, and individuals. It is inherently diverse, resilient, and deeply rooted in local traditions.

Implementing agricultural biodiversity begins with conserving genetic resources that will be vital for future food production. Genetic resources are an irreplaceable asset for humanity's future. Promoting their conservation and sustainable use through international cooperation, while ensuring fair benefit-sharing with countries of origin, is critical for sustainable food production. To achieve this, it is essential to secure genetic resources and strengthen collaborative relationships with research institutions and universities worldwide. Programs such as SATREPS (Science and Technology Research Partnership for Sustainable Development) play a key role in advancing research and conservation efforts in biological resources.

References

FAO (2010) The second report on the state of the world's, plant genetic resources for food and agriculture. Commission on Genetic Resources for Food and Agriculture, Food and Agriculture Organization of the United Nations, Rome

Johns T, Francisca I, Eyzaguirre PB (2006) Agrobiodiversity, nutrition, and health: understanding the links between agriculture and health. ResearchGate. International Food Policy Research Institute USA. https://www.researchgate.net/publication/5055727

Keding GB, Cogill B (2013) Linking nutrition and agrobiodiversity. ICN2 second international conference on nutrition better nutrition better lives. FAO and WHO. https://www.fao.org/3/as567e/as567e.pdf

MIX
Papier aus verantwortungsvollen Quellen
Paper from responsible sources
FSC® C105338

If you have any concerns about our products,
you can contact us on
ProductSafety@springernature.com

In case Publisher is established outside the EU,
the EU authorized representative is:
Springer Nature Customer Service Center GmbH
Europaplatz 3, 69115 Heidelberg, Germany

Printed by Libri Plureos GmbH
in Hamburg, Germany